Regine Frank

Light Propagation and Random Lasing

Regine Frank

Light Propagation and Random Lasing

Microscopic Theory of Random Lasing and Light Transport in Amplifying Disordered Media

Südwestdeutscher Verlag für Hochschulschriften

Impressum/Imprint (nur für Deutschland/only for Germany)
Bibliografische Information der Deutschen Nationalbibliothek: Die Deutsche Nationalbibliothek verzeichnet diese Publikation in der Deutschen Nationalbibliografie; detaillierte bibliografische Daten sind im Internet über http://dnb.d-nb.de abrufbar.
Alle in diesem Buch genannten Marken und Produktnamen unterliegen warenzeichen-, marken- oder patentrechtlichem Schutz bzw. sind Warenzeichen oder eingetragene Warenzeichen der jeweiligen Inhaber. Die Wiedergabe von Marken, Produktnamen, Gebrauchsnamen, Handelsnamen, Warenbezeichnungen u.s.w. in diesem Werk berechtigt auch ohne besondere Kennzeichnung nicht zu der Annahme, dass solche Namen im Sinne der Warenzeichen- und Markenschutzgesetzgebung als frei zu betrachten wären und daher von jedermann benutzt werden dürften.

Verlag: Südwestdeutscher Verlag für Hochschulschriften GmbH & Co. KG
Heinrich-Böcking-Str. 6-8, 66121 Saarbrücken, Deutschland
Telefon +49 681 37 20 271-1, Telefax +49 681 37 20 271-0
Email: info@svh-verlag.de

Approved by: Bonn, Uni, Diss, 2011

Herstellung in Deutschland:
Schaltungsdienst Lange o.H.G., Berlin
Books on Demand GmbH, Norderstedt
Reha GmbH, Saarbrücken
Amazon Distribution GmbH, Leipzig
ISBN: 978-3-8381-3009-5

Imprint (only for USA, GB)
Bibliographic information published by the Deutsche Nationalbibliothek: The Deutsche Nationalbibliothek lists this publication in the Deutsche Nationalbibliografie; detailed bibliographic data are available in the Internet at http://dnb.d-nb.de.
Any brand names and product names mentioned in this book are subject to trademark, brand or patent protection and are trademarks or registered trademarks of their respective holders. The use of brand names, product names, common names, trade names, product descriptions etc. even without a particular marking in this works is in no way to be construed to mean that such names may be regarded as unrestricted in respect of trademark and brand protection legislation and could thus be used by anyone.

Publisher: Südwestdeutscher Verlag für Hochschulschriften GmbH & Co. KG
Heinrich-Böcking-Str. 6-8, 66121 Saarbrücken, Germany
Phone +49 681 37 20 271-1, Fax +49 681 37 20 271-0
Email: info@svh-verlag.de

Printed in the U.S.A.
Printed in the U.K. by (see last page)
ISBN: 978-3-8381-3009-5

Copyright © 2012 by the author and Südwestdeutscher Verlag für Hochschulschriften GmbH & Co. KG and licensors
All rights reserved. Saarbrücken 2012

Microscopic Theory of Random Lasing and Light Transport in Amplifying Disordered Media

Dissertation

zur

Erlangung des Doktorgrades (Dr. rer. nat.)

der

Mathematisch-Naturwissenschaftlichen Fakultät

der

Rheinischen Friedrich-Wilhelms-Universität Bonn

vorgelegt von

Regine Frank

aus Geislingen an der Steige

Bonn 2009

Angefertigt mit Genehmigung der Mathematisch-Naturwissenschaftlichen Fakultät der Rheinischen Friedrich-Wilhelms-Universität Bonn.

1.Referent: Prof. Dr. Johann Kroha
2.Referent: Prof. Dr. Kurt Busch

Tag der Promotion: 26. April 2010
Erscheinungsjahr: 2011

Abstract

In the last decade Anderson Localization of Light and Random Lasing has attracted a variety of interest in the community of condensed matter theory. The fact that a system so inartificial as a thin layer consisting of dust particles, which have amplifying properties, can produce coherent laser emission, is of a simple elegance, which promises new insights in fundamental physics. The Random Laser consists of randomly distributed scatterers which have amplifying properties, embedded in an either amplifying or a passive medium. There is no need of an external feedback mechanism, like in the system of a conventional laser. Despite the striking chasteness of the effect, there are still vivid discussions about theory for the random laser, which has not been fully described yet. Whereas there are several attempts to enter the subject numerically by considering cavity approximations, this thesis is concerned with building a microscopically self-consistent theory of random lasing. In order to design this method we had to study light localization effects in random media including absorption and gain. We incorporated interference effects by calculating the Cooperon contributions and we found, that we reach Anderson localized states for passive media. Mapping this theory on a system which consists of laser active Mie-scatterers in a passive medium, we found that by incorporation of the Cooperon, we loose the Anderson localization again, but the system is still weakly localized, which is sufficient for the enhancement of population inversion and stimulated emission. The description of a random lasing system is completed by coupling the analytically derived microscopic transport theory to the laser rate equations of a four level laser and solving the system numerically self-consistent. Finally we develop a generalized method for describing light transport, multiple scattering and interference effects in a translationally non-invariant system of finite size analytically. The solution of this theory coupled to the rate equations gives us a closed theory to calculate transport and lasing properties of a random lasing slab geometry.

Contents

1 Introduction — 7

2 Light Waves in Random Media — 11
- 2.1 Maxwell Equations — 11
- 2.2 Green's Function Formalism — 13
  - 2.2.1 Single Particle Green's Function — 13
  - 2.2.2 Two Particle Green's Function — 13
  - 2.2.3 Higher Moments of the Green's Function — 14
- 2.3 Scattering Formalism — 14
  - 2.3.1 The T Operator — 15
  - 2.3.2 Configurational Averaging — 15
  - 2.3.3 The Self-Energy — 16
- 2.4 Two Particle Quantities and Bethe Salpeter-Equation — 16
  - 2.4.1 Incoherent and Coherent Contributions — 18

3 Light Transport in Infinite Three Dimensional Disordered Media with Absorption or Amplification — 21
- 3.1 Introduction — 21
- 3.2 Light Matter Interaction — 22
- 3.3 Setup and Model — 24
- 3.4 Light Propagation in Ladder Approximation — 25
- 3.5 Theory of Transport and Localization — 28
  - 3.5.1 Expansion of Two-particle Green's Function into Moments — 29
  - 3.5.2 Computing the Coefficients of the Expansion into Moments — 29
  - 3.5.3 General Solution of the Bethe-Salpeter Equation — 30
  - 3.5.4 Vertex Function and Self-Consistency — 32
- 3.6 Results and Discussion of the Transport Theory in Infinite 3-D Media with Absorption and Gain — 32
  - 3.6.1 Discussion Transport Theory — 33
  - 3.6.2 Causality and Length Scales — 35
- 3.7 Conclusion — 36

4 Phenomenological Random Laser — 39
- 4.1 Introduction — 39
- 4.2 Theory of a Diffusive Random Laser — 40
- 4.3 Results and Discussion — 42
- 4.4 Summary — 44

5 Selfconsistent Microscopic Theory of Random Lasing — 45
- 5.1 Single - Particle Green's Function — 45
- 5.2 Light - Intensity Correlation Function — 48
- 5.3 Fourier - Transformed Bethe - Salpeter Equation — 51
- 5.4 Light Intensity Transport in Bounded Disordered Media with Absorption or Gain — 52
  - 5.4.1 Solution by Moment Expansion — 52

		5.4.2	Continuity Equation	53
		5.4.3	Current Relaxation Equation	54
	5.5	Diffusion Pole Structure		55
		5.5.1	Selfconsistent Diffusion Coefficient	58
	5.6	Microscopic Theory of Random Lasers		58
		5.6.1	Microscopic Determination of the Amplification Rate of the Intensity	59
		5.6.2	Numerics Procedure	61
	5.7	Numerical Results and Discussion		62
		5.7.1	Film of ZnO at 50 % Filling	62
		5.7.2	Behavior of the Correlation Length for Various System Parameters	70
		5.7.3	Behavior of the Diffusion Coefficient for Various System Parameters	75
		5.7.4	Different Filling Fractions	78
	5.8	Conclusion		82

6 Summary — 83

Appendices — 84

A Dyson Equations — 85

B Disorder Averaged Full Single-Particle Green's Function — 87
  B.1 Analytic Calculation of $G(\vec{r},\vec{r}')$ — 88

**C Two Particle Green's Function -
The Intensity Correlator** — 93

D Calculating the Memory Kernel $M(\Omega)$ — 97

E Technical Transformations — 101
  E.1 Transformation of the Coordinate System — 101
  E.2 Transformation used for Single Particle Greens Function — 101

Bibliography — 103

Chapter 1

Introduction

This thesis is concerned with light propagation, transport and amplification in random media in general, which is a wide area of research. Whereas we find undamped propagation in the case of vacuum, in the random medium we have to consider mainly two different mechanisms of transport. First the diffusive processes which lead to an exponential decay of the light intensity and second the interference processes which may lead to Anderson localization. Finally the interplay of these transport mechanisms may lead in amplifying media to an effect which is called random lasing.

Figure 1.1: *Schematic illustration of light transport in the three different regimes. Shown is the incident light pulse (intensity) in position space (upper row) at $t = 0$ and the same light pulse after a finite time $t > 0$. In the ballistic regime, i.e. in the absence of disorder (left column), light propagates with the bare speed of light, whereas in diffusive regime with weak disorder (center column), the initial pulse decays slower. Finally in the localized regime, i.e. strong disorder, (right column) the light pulse does not decay at all, light intensity is localized within a given volume.*

Random lasing is a phenomenon which has been discussed during the last decade for various disordered media. Optical gain in such a laser is either achieved by introducing a laser dye, or the scattering structure is excited to deliver optical gain itself [1, 2]. In the first case passive scatterers are embedded in an optically active host medium, whereas in the latter case the host medium is passive and the scatterers are themselves active. A combination of those two realizations should also be accessible. One example of a scattering and amplifying random media is ZnO powder.
ZnO powders are ideally suited for application in random lasing because of their relatively high refractive index (ZnO the refractive index is 2.3 in the ultraviolet near-band-edge region) leading to strong light scattering in combination with an adequate optical gain if sufficiently excited [3, 4] .

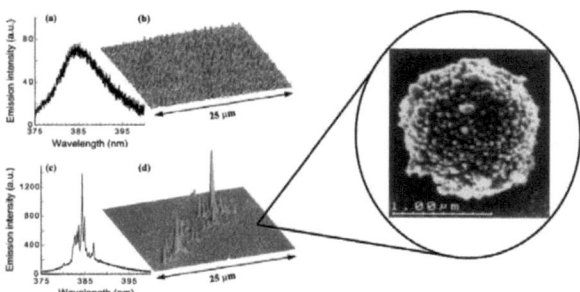

Figure 1.2: *The experimental results with ZnO semiconductor powder show sharp spectral features and lasing emission from spatially confined regions. H. Cao et al., PRL* **84***, 5584 (2000)*

A conventional laser is usually constructed from two basic elements: a material that provides optical gain through stimulated emission and an optical cavity that partially traps the light and therefore provides the necessary feedback mechanism. When the total gain in the cavity is larger than the losses, the system reaches a threshold and lases. It is the cavity that determines the modes of a laser, that means, it determines not only the directionality of the output but also its frequency. Random lasers work on the same principles, but the modes are determined by multiple scattering as intrinsic feedback and not by a laser cavity.

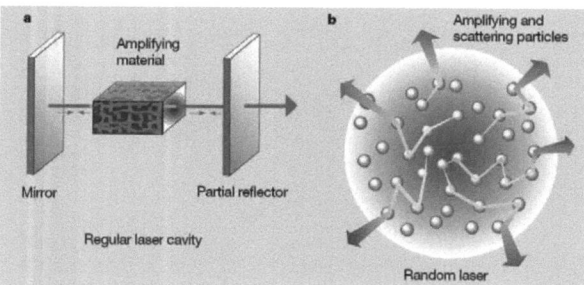

Figure 1.3: *D.S. Wiersma et al., Nature* **406***, 132 (2000) (a) Conventional Laser versus (b) Random Laser. Whereas the conventional laser consists of an amplifying material and an external feedback mechanism, the Random Laser contains an intrinsic feedback mechanism, all the multiple scattering processes, and therefore the gain is intrinsically influenced.*

Multiple scattering as the underlying principle of light transport in disordered media occurs in nearly all optical materials that appear opaque and is very common in nature. In a lot of processes multiple scattering is the origin of loss of intensity, so is mainly a process which has been tried to avoid.
This type of propagation (see Fig. 1.2) is that of a random walk, just as in the Brownian motion of particles suspended in a liquid. The fundamental parameters describing multiple scattering are the mean free path (the average step size in the random walk) and the diffusion constant. Scattering in disordered optical materials is complex yet completely coherent. This means that the phase of each

of the optical wavelets undergoing a random walk is well defined and interference effects can occur, even if a material is strongly disordered.
The most self-evident visualization of interference of multiply scattered light is that of a laser speckle, which is the grainy pattern observed when looking at a laser pointer that is scattered from a piece of paper. As interference we can discuss two or more wave interference as well as self interference. It is important to remark, that interference contributions arise only due to light paths in the sample which are identical but traveled along in opposite directions, these are the so-called time-reversed paths as illustrated in Fig 1.4. As a consequence we find that in the direction of back reflection the light intensity is amplified by a factor of two, only due to the existence of such interference phenomena. Experimentally such phenomenon is observed and named Coherent backscattering effect.

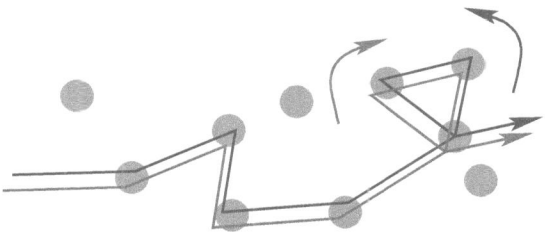

Figure 1.4: *Inverse path-directions lead to interference effects.*

The difference between light diffusion and multiple scattering is that (bare) diffusion refers to a simplified picture of multiple scattering in which interference effects are neglected. Multiple scattering due to randomness not only occurs in natural materials, but is also intrinsically present in photonic materials, such as photonic crystals, intended for the realization of optical devices. In those materials, multiple scattering has always been considered an unwanted property arising from structural artefacts. It has now become clear that such artefacts are difficult to avoid [5]. Using multiple scattering to introduce new functionalities therefore opens up a completely new perspective on disorder in photonic materials.
An additional effect requiring consideration regarding light transport is the fact that the randomly distributed scatterers are of finite size. This leads to geometrical resonances within the scatterers itself. In case of spherical scatterers such resonances are called Mie resonances. In two dimensional structures this effect is also called a whispering gallery mode. If the light frequency is in or close to a Mie resonance, then the transport of intensity is affected, because light can be stored in such resonances, and therefore the transport is slowed down, resulting in a smaller diffusion constant.
To develop a theory that can describe all aspects of a random laser is very difficult. A complete model would have to include the dynamics of the gain mechanism because gain saturation forms an intrinsic aspect of an amplifying system above threshold. Without gain saturation, the intensity would diverge leading to unphysical results. In addition, interference effects have to be included to describe the mode structure. Interference in multiple scattering leads to a granular distribution of the intensity called speckle (Fig. 1.2). In most random materials, the intensity is spread throughout the sample and the modes are extended. In certain random materials, interference can lead to an effect called light localization [6, 7, 8], which is the optical counterpart of Anderson localization of electrons [9]. Owing to interference, the free propagation of waves and thereby the multiple scattering process, comes practically to a halt in that case. This can be understood in terms of the formation of randomly shaped but closed modes with an overall exponentially decaying amplitude (Fig. 1.2). The average spatial extent of these localized modes defines a length scale called the localization length. Localization can only take place in optical materials that are extremely strongly scattering, the requirement being that the mean free path becomes smaller than the reciprocal wave vector : $kl < 1$. This is also known as the Ioffe - Regel criterion [10]. The need for a detailed model of random lasing became clear after an observation by Cao and co-workers, who found that

carefully performed experiments revealed narrow spikes in the emission spectrum on top of a global narrowing [11]. Attempts to understand the origin of these spikes have led to a vivid discussion in the literature.

In a random laser, light is confined within an amplifying disordered medium due to permanent scattering on a microscopic length scale. The light is thus retarded from leaving the active region up to the limit of localized light [15, 16]. The localization of light is caused by constructive interference, because elastic scattering of light is a completely coherent process.

The onset of lasing is demonstrated typically by a threshold-like behavior of the optical emission [4], but only under certain conditions do narrow laser lines, similar to those of conventional lasers, occur [12]. The origin of such spectrally narrow laser lines is still open to interpretation. It has been attributed to strongly localized modes of the light in the sense of an Anderson localization [13] that cannot be explained in a purely diffusive model where interference effects are neglected [14].

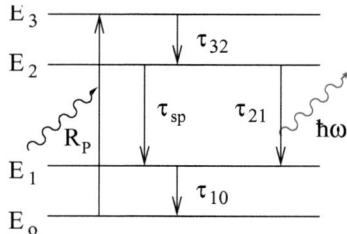

Figure 1.5: *Schematic description of a conventional four level laser. The reservoir electrons (ground level) are pumped to the third level, from where they relax almost immediately to the second level. Finally an inversion between the second and the first level is reached and by stimulated emission the electrons relax coherently into the first level. This process is called lasing. Finally they relax to the ground level.*

In this thesis there is a transport theory described which has been developed for either amplifying or absorbing bulk material with embedded amplifying or absorbing Mie scatterers in random positions. The theory is further expanded for samples with finite sizes and also coupled to a system of rate equations for a conventional four level laser (see Fig. 1.5). This system is solved self-consistently.

Chapter 2

Light Waves in Random Media

In this chapter we want to introduce briefly the various well known methods and strategies used in this theses to derive a self-consistent theory of light localization and random lasing in disordered media. The given introduction is rather general than complete and the interested reader is asked to proceed to the references given here. Nevertheless the introduction to Maxwell's equations, the Method of Green's functions and the information provided by the diverse moments are very important for the translations of the physical pictures into the theory of this thesis.

2.1 Maxwell Equations

The original, microscopic Maxwell-Theory of electromagnetism has a very clear and simple structure which is the reason for its beauty. Electromagnetic phenomena such as light are described by Maxwell's equations, which determine the evolution of the electromagnetic fields. These are vector-valued functions of space and time and known as the electric field $\vec{E}$ and the magnetic induction $\vec{B}$. Further we introduce the charge and current densities ρ and $\vec{J}$ which are the sources for these fields. If the charges and currents are known, then the evolution of the fields $\vec{E}$ and $\vec{B}$ is governed by the following coupled set of partial differential equations, the microscopic equations in vacuum expressed in Heaviside-Lorentz units [66]

$$\partial_t \vec{E} = c\vec{\nabla} \times \vec{B} - \vec{J} \quad (2.1)$$
$$\partial_t \vec{B} = -c\vec{\nabla} \times \vec{E} \quad (2.2)$$
$$\vec{\nabla} \cdot \vec{E} = \rho \quad (2.3)$$
$$\vec{\nabla} \cdot \vec{B} = 0. \quad (2.4)$$

We have chosen the system of units in which all quantities are dimensionless, and the electric permittivity ϵ_{vac} and the magnetic permeability μ_{vac} are equal to 1. If we place some material in the vacuum, the atoms comprising it will be affected by the electromagnetic fields and via their motion produce (excess) charges and currents. The response of the atoms is in general determined from quantum mechanical calculations, and the coupled material-Maxwell system would have to be solved self-consistently. This is practically impossible and in addition unnecessary, since the variation of the fields on the atomic lengths scales is not interesting. Therefore some kind of simplification has to be introduced in the way that we establish the so called macroscopic fields. In order to obtain these macroscopic equations, we separate charge and current density into two parts $\rho = \rho_1 + \rho_2$, $\vec{J} = \vec{J}_1 + \vec{J}_2$, where the ones describe the free parts and the others the bound parts Ref.[67]. To eliminate ρ_2 and $\vec{J}_2$ the polarization and the magnetization of the sample are introduced

$$\rho_2 = -\vec{\nabla} \cdot \vec{P} \quad \vec{J}_2 = -(\partial_t \vec{P} + c\vec{\nabla} \times \vec{M}) \quad \text{with} \quad \dot{\rho}_2 + \vec{\nabla} \cdot \vec{J}_2 = 0 \quad (2.5)$$

This is always possible but not unique. Introducing two additional new fields

$$\vec{H} \equiv \vec{B} - \vec{M} = \vec{B} - \chi_m \vec{B} = \frac{1}{\mu}\vec{B} \qquad (2.6)$$

$$\vec{D} \equiv \vec{E} + \vec{P} = \vec{E} + \chi_e \vec{E} = \epsilon \vec{E} \qquad (2.7)$$

we eliminate $\vec{P}$ and $\vec{M}$, where the rightmost equalization in Eq. (2.6) and Eq. (2.7) is valid in a linear response approximation. The magnetic field H and the electric displacement D describe the electro-magnetic fields within matter, and the susceptibility χ_e and χ_m characterize the response of the system to external fields, and they are therefore material dependent properties which can be calculated microscopically.
This procedure leads us finally to the macroscopic Maxwell's equations of electrodynamics

$$\partial_t \vec{D} = c\vec{\nabla} \times \vec{H} + \vec{J} \qquad (2.8)$$
$$\partial_t \vec{B} = -c\vec{\nabla} \times \vec{E} \qquad (2.9)$$
$$\vec{\nabla} \cdot \vec{D} = \rho \qquad (2.10)$$
$$\vec{\nabla} \cdot \vec{B} = 0 \qquad (2.11)$$

By performing a coarse-grained average, which leaves the linear relations above unchanged, one finds the macroscopic interpretation: $\vec{E}$ and $\vec{B}$ are the averaged microscopic fields. The fields $\vec{H}$ and $\vec{D}$ can be understood by identifying $\vec{P}$ and $\vec{M}$ in leading order with the electric and magnetic dipole moment densities.
The dielectric constant is a macroscopic quantity, which represents on average the linear response of matter to an external field. By assuming ϵ to be complex, we introduce dissipation. We find absorption of light in case of a positive sign of the imaginary part of ϵ, and amplification of light for a negative sign. The macroscopic Maxwell's equations Eqs. (2.8-2.11) together with the two linear constitutive relations Eq. (2.6) and Eq. (2.7) represent a complete and unique system of equations. For a more detailed discussion see e.g. [67].
It is convenient to eliminate either the electric or the magnetic field in one of the dynamic Maxwell's equations in order to obtain a closed equation in one of the fields. The vector wave equations

$$\vec{\nabla} \times \vec{\nabla} \times \vec{E}(\vec{r},t) + \epsilon(\vec{r})\frac{\partial^2}{\partial t^2}\vec{E}(\vec{r},t) = -\frac{\partial}{\partial t}\vec{J}(\vec{r},t) \qquad (2.12)$$

$$\vec{\nabla} \times \frac{1}{\epsilon(\vec{r})}\vec{\nabla}\vec{B}(\vec{r},t) + \frac{\partial^2}{\partial t^2}\vec{B}(\vec{r},t) = \vec{\nabla} \times \frac{1}{\epsilon(\vec{r})}\vec{J}(\vec{r},t). \qquad (2.13)$$

Changing from time to frequency space by performing a standard Fourier transformation, Eq. (2.12) reads as

$$\vec{\nabla} \times \vec{\nabla} \times \vec{E}(\vec{r},\omega) - \omega^2 \epsilon(\vec{r})\vec{E}(\vec{r},\omega) = i\omega \vec{J}(\vec{r},\omega) \qquad (2.14)$$

Now we use the vector identity $\vec{\nabla} \times \vec{\nabla} \times \vec{v} = \vec{\nabla}(\vec{\nabla} \cdot \vec{v}) - \vec{\nabla}^2 \vec{v}$ for an arbitrary vector field $\vec{v}$, and derive the scalar wave equation by neglecting polarization and therefore neglecting the term $\vec{\nabla}(\vec{\nabla} \cdot \vec{v})$

$$-\vec{\nabla}^2 E(\vec{r},\omega) - \omega^2 \epsilon(\vec{r}) E(\vec{r},\omega) = j(\vec{r},\omega). \qquad (2.15)$$

With the scalar wave equation we have reached the description of propagation of an electric field, which is the basis on which we build the theory of propagation of light intensity in disordered media.

2.2 Green's Function Formalism

Before going into more detailed considerations in chapter 3, however, the discussion of the overall approach of Green's functions and the discussion of the physical meaning of the several moments is useful.

In a random medium, a complete solution of the wave equation is represented by the knowledge of the Green's function $G(\omega, r, r')$ for all values of ω, r, r' in the presence of disorder $\sigma(r)$. The accurate solution of the wave equation is not generally possible in the presence of $\sigma(r)$. There is also the problem of how to extract the desired information from $G(\omega, r, r')$ even if it were known.

At this point a look at the random walk problem compared to the propagation of light is helpful. Assume that the position of a random walker is known at every instant of time, given by $r(t)$. This is equivalent to the determination of the Green's function $G(\omega, r, r')$ for the wave equation and represents the complete solution of the random walk problem in principle. The diffusive behavior, however, which results from the statistical character of $r(t)$, is not immediately obvious from $r(t)$. It is contained within the evaluation of the moments of $r(t)$, i.e. $\langle r \rangle_c$, $\langle r^2 \rangle_c$ etc., where the angular brackets with the subscript c denote averaging over different configurations, the random perturbations $\sigma(r)$ in the wave scattering case, and different random walk trajectories in the random walker case. When we look at the transport of light in disordered media the first objective of our approach is the approximate evaluation of $\langle G \rangle_c$, since its calculation is simpler than that for G, but we also pay the price of loosing some information. For example, whereas G depends on the source position r', $\langle G \rangle_c$ depends only on the source-detector separation $r - r'$, because after configurational averaging, $\langle G \rangle_c$ can no longer depend on any particular $\sigma(r)$, and so only $r - r'$ plus the statistical properties of the $\sigma(r)$ ensemble are relevant. In other words, an averaged quantity cannot depend on the averaged variable, and in this case independence from $\sigma(r)$ means all spatial positions are equivalent.

For the work presented in this thesis the knowledge of several moments of the Green's function is important.

2.2.1 Single Particle Green's Function

The disorder averaged Green's function $\langle G \rangle_c$, the first moment of G, shows wave propagation characteristics in an averaged sense. Metaphorically spoken, this is the "effective medium" as it is seen by a wave. Additionally, it marks the spatial scale, where beyond this effective medium description can no longer be valid. Mathematically, this point is expressed in the exponential decay of $\langle G \rangle_c$ as a function of $|r - r'|$. Yet this decay does not imply wave localization at all, just as in random walk $\langle r \rangle_c = 0$ does not indicate that the random walker is localized at the origin. The exponential decay of $\langle G \rangle_c$ means that the wave coherence, which means unique wave propagation direction and phase relation, is lost. This decay length is defined as the mean free path. For the random walk, $\langle r \rangle_c = 0$ implies the absence of ballistic motion.

2.2.2 Two Particle Green's Function

Whereas $\langle G \rangle_c$ shows the character of the coherent part of the wave propagation in a random medium, all the transport effects beyond the scale of the mean free path are included in the second moment, the disorder-averaged Green's function $\langle GG^* \rangle_c$, just as in the random walk case the diffusion dynamics emerge from the time dependence of $\langle r^2 \rangle_c$. Therefore the next objective in our approach is the evaluation of $\langle GG^* \rangle_c$. The result of this calculation will display that at the long-time, long-distance limit, intensity transport is indeed diffusive in character, and the associated diffusion constant will be explicitly calculated in this thesis. In the context of diffusive transport the self-interference effect can be demonstrated to strongly influence the diffusion constant. Since one of the main objects of this thesis is to describe light intensity transport in disordered media, we focus on the evaluation and the characteristics of two particle Green's functions.

2.2.3 Higher Moments of the Green's Function

As our main task is to calculate the intensity transport behavior, which is given by $\langle GG^* \rangle_c$, we do not consider fluctuations in the intensity transport, which have to be calculated from the fourth moment $\langle GG^*GG^* \rangle_c$. This is the intensity- intensity correlation and it contains information about the long-range memory and phase interference effects that may be retrieved from intensity- or current-fluctuations, e.g., speckle patterns formed by light after passing through a random medium.
The set of these moments of G provide the insight into the special statistical character of wave transport in random media. Wave transport characteristics obtained from the moments of G necessarily differ from those in a single configuration. Nevertheless it is important to outline that the moments can give a general understanding of the single-configuration characteristics because in most physical situations the ergodic hypothesis is valid, which means that the configurationally averaged behavior may be equated to the infinite-time average of the single-configuration behavior.

2.3 Scattering Formalism

As outlined before the Green's function formalism is a very useful tool to gain insight to the physical behavior of a wave in disordered media and its statistical characteristics. Especially, it can solve inhomogeneous differential equations. To make the following considerations a little better to read we change the notation according to $\tilde{G} = G$ as the non averaged Green's function and $G = \langle G \rangle_c$ as the disorder averaged Green's function.
For a system which is exited at $t = 0$ by a light pulse, the vector wave equation (Helmholtz equation), cf. Eq. (2.14), reads

$$-\vec{\nabla} \times \vec{\nabla} \times \tilde{\boldsymbol{G}}(\vec{r}, \vec{r}', \omega) + \omega^2 \epsilon(\vec{r}) \tilde{\boldsymbol{G}}(\vec{r}, \vec{r}', \omega) = \delta(\vec{r} - \vec{r}') \mathbb{1}, \qquad (2.16)$$

$\mathbb{1}$ is the unity operator in three dimensional space. We introduce $g(\omega)$ and $h(\omega)$ and rewrite Eq. (2.16)

$$g(\omega) := \frac{\omega^2}{c^2} \epsilon_0(\omega) \quad \text{and} \quad h(\omega) := -\frac{\omega^2}{c^2} \left[\epsilon_{scat}(\omega) - \epsilon_0(\omega)\right] \qquad (2.17)$$

$$-\vec{\nabla} \times \vec{\nabla} \times \tilde{\boldsymbol{G}}(\vec{r}, \vec{r}', \omega) + [g(\omega) - h(\omega) V(\mathbf{r})] \tilde{\boldsymbol{G}}(\vec{r}, \vec{r}', \omega) = \delta(\vec{r} - \vec{r}') \mathbb{1}, \qquad (2.18)$$

The dielectric function is expressed as a function of the background contribution ϵ_0 and a spatial dependent contribution ϵ_{scat} of the scatterers

$$\epsilon(\mathbf{r}, \omega) = \epsilon_0(\omega) + [\epsilon_{scat}(\omega) - \epsilon_0(\omega)] V(\mathbf{r}) \quad \text{and} \quad \epsilon(\mathbf{r}, \omega) \in \mathbb{C} \qquad (2.19)$$

where $V(\mathbf{r})$ is the scatterers volume function. This transformation allows us to write the electromagnetic potential see Eq. (2.18) in a more intuitive form, where the function $g(\omega)$ describes the undisturbed background medium. In the electronic case this is expressed by the free propagator in a homogenous medium. Henceforth, $h(\omega)$ characterizes the influence of the scatterers. Again in electronic language this would lead to the full Green's function including potential scattering. In the following procedure we use this notation of free and full Green's function.
From Eq. (2.16) we obtain the relation between the electric field and the Green's function

$$\vec{E}(\vec{r}, \omega) = \int \mathrm{d}^3\vec{r}' \, \tilde{\boldsymbol{G}}(\vec{r}, \vec{r}'; \omega) \vec{j}(\vec{r}'; \omega) \qquad (2.20)$$

which displays that the intrinsic information of the system is fully included in the Green's function. Green's functions are the generic building block of the multiple scattering theory. If the Green's function for a given problem is derived, which can be of course a difficult task, one can find a formal Born series expansion in the scattering potential.

2.3. SCATTERING FORMALISM

$$G = G_0 + G_0 V G \tag{2.21}$$

2.3.1 The T Operator

By iterating this self-consistent equation Eq. (2.21), we derive an alternative way to express the Green's function

$$\begin{align}
G &= G_0 + G_0 V [G_0 + G_0 V G] \tag{2.22} \\
&= G_0 + G_0 V G_0 + G_0 V G_0 V G_0 + \ldots \tag{2.23}
\end{align}$$

Eq. (2.23) expresses clearly that each scattering event (V) is a source for a further propagating wave

$$\begin{align}
G &= G_0 + G_0 T G_0 \tag{2.24} \\
&= G_0 + G_0 V G_0 + G_0 V G_0 V G_0 + \ldots \tag{2.25}
\end{align}$$

where

$$\begin{align}
T &= V + V G_0 V + V G_0 V G_0 V + \ldots \tag{2.26} \\
&= V (\mathbb{1} - G_0 V)^{-1} \tag{2.27}
\end{align}$$

is called the scattering matrix T. From the defining equation, Eq. (2.26), it is seen that if V represents one scattering event, than T includes all multiple scattering events [71]. In general Eq. (2.26) is written as

$$G^{-1} = G_0^{-1} - V. \tag{2.28}$$

The solution of the general wave equation

$$[\nabla^2 + \epsilon_0 \omega^2 - \sigma(r)]\phi(\omega, r) = 0 \tag{2.29}$$

$$\left(\nabla^2 + \epsilon_0 \omega^2\right) \phi(\omega, r) = \sigma(r)\phi(\omega, r). \tag{2.30}$$

can be completely expressed with the help of the $\mathbf{T}$ matrix in terms of the uniform-medium solutions $\phi_0(\omega, r)$ and $G_0(\omega, r - r')$. Then the solution may be written in Dirac's notation

$$\begin{align}
|\phi\rangle &= |\phi_0\rangle + \mathbf{G}_0 \mathbf{V} |\phi\rangle = |\phi_0\rangle + \mathbf{G}_0 \mathbf{V} |\phi_0\rangle + \mathbf{G}_0 \mathbf{V} \mathbf{G}_0 \mathbf{V} |\phi_0\rangle + \ldots \tag{2.31} \\
&= |\phi_0\rangle + \mathbf{G}_0^+ \mathbf{T}^+ |\phi_0\rangle. \tag{2.32}
\end{align}$$

The + superscripts on $\mathbf{G}_0$ and $\mathbf{T}$ in the last line of Eq. (2.31) are meant to select the physical solution branch where the scattering from an inhomogeneity is represented by an outgoing wave (from the inhomogeneity) rather than by an incoming wave, which would be selected by $\mathbf{G}_0^- \mathbf{T}^-$.

2.3.2 Configurational Averaging

From Eq. (2.24), the configurationally averaged Green function is given by

$$\langle \mathbf{G} \rangle_c = \mathbf{G}_0 + \mathbf{G}_0 \langle \mathbf{T} \rangle_c \mathbf{G}_0, \tag{2.33}$$

where $\mathbf{G}_0$ is independent of $\sigma(r)$ and is therefore not affected by the average. From the T representation of Eq. (2.26), we have

$$\langle \mathbf{T} \rangle_c = \langle \mathbf{V}(1 - \mathbf{G}_0 \mathbf{V})^{-1} \rangle_c. \qquad (2.34)$$

$\langle T \rangle_c$ contains all the higher-order correlations of V as we have seen above, such as $\langle VG_oV \rangle_c$ and $\langle VG_oVG_oV \rangle_c$, and all the multiple scattering caused by V. In the real space representation $\langle G \rangle_c$ depends only on the spatial separation between the source and the receiver $r - r'$. Therefore its Fourier transform is a function of one k only, just as for G_0. It can be shown that a convolution integral in real space, as symbolized by $G_0 \langle T \rangle_c G_0$, means simple multiplication of the transformed quantities in k space. Therefore it is obvious that in the k representation, $\langle T \rangle_c$ is also a function of one k only, which implies a dependence on only the separation of source and destination $r - r'$ in real space.

2.3.3 The Self-Energy

Equivalently to the derivation of Eq. (2.28), from Eq. (2.21) one can define a self energy Σ operator as

$$\langle \mathbf{G} \rangle_c^{-1} = \mathbf{G}_0^{-1} - \Sigma, \qquad (2.35)$$

In the k representation, Σ is also a function of one k only, because both $\langle G \rangle_c$ and G_0 are only functions of one k. This causes a dependence on $r - r'$ in real space, just as for $\langle T \rangle_c$. From (3.12) and (3.14), Σ is related to $\langle T \rangle_c$ by

$$\langle \mathbf{\Sigma} \rangle_c = \langle \mathbf{T} \rangle_c (1 - \langle \mathbf{T} \rangle_c \mathbf{G}_0)^{-1}. \qquad (2.36)$$

Σ is denoted the self-energy operator, and Eq. (2.35) is known as the Dyson equation. In momentum space, both Eq. (2.35) and Eq. (2.36) are simple algebraic equations. All the relevant operators are diagonal matrices. The comparison of Eq. (2.35) with Eq. (2.28) shows that in spite of the apparent similarity, the self-energy Σ is a very different object from the operator V of the exact Green function in a fixed configuration. In position space the operator V represents scattering-off of a local perturbation potential $\sigma(r)$. However Σ is a nonlocal operator as can be seen from the defining equation for $\langle G \rangle_c$. In momentum space is given by

$$\left[\epsilon_0 \omega^2 - k^2 - \Sigma(k,\omega) \right] \langle \mathbf{G} \rangle_c = 1 \qquad (2.37)$$

Applying a Fourier-Transformation to position space, i.e. replacing $-k^2$ by ∇^2, and additionally using that a multiplication in the k-domain means convolution in the real space, so that the above equation now reads

$$\left[\epsilon_0 \omega^2 + \nabla^2 \right] \langle \mathbf{G} \rangle_c (\omega, r - r') \qquad (2.38)$$
$$- \int dr_1 \Sigma(\omega, r - r_1) \langle \mathbf{G} \rangle_c (\omega, r_1 - r') = \delta(r - r')$$

This equation reduces to the ordinary wave equation only if $\Sigma(\omega, r - r_1) = \Sigma(\omega)\delta(r - r_1)$. This turns out to be possible when the effective medium description is valid.

2.4 Two Particle Quantities and Bethe Salpeter-Equation

Since the disorder averaged Green's function is translational invariant, it is not suitable for transport properties of the random system, since they represent collective phenomena. This means that the

2.4. TWO PARTICLE QUANTITIES AND BETHE SALPETER-EQUATION

physical quantity which is transported through the random system is the energy, which is a conserved quantity. On a purely phenomenological level, we may say that the restriction imposed by the conservation of energy leads to a correlation between the motion of the field quanta. This process cannot be described by averaged single particle quantities alone. Actually it has to be expressed in terms of the averaged product of the two non-averaged single particle Green's functions, since the energy density is the product of the two amplitudes.

The conservation of energy is expressed by the continuity equation for the energy density $\Phi(\vec{r},t)$, and the energy current density $\vec{J}(\vec{r},t)$

$$\frac{d\Phi}{dt} + \nabla \cdot J = \delta^4(t,\vec{r}) \tag{2.39}$$

The Fourier transformed quantities of the energy density Φ and the energy current density J are connected by the energy transport velocity v_E. The energy current density can be expressed as the gradient of the energy density assumed that the system characteristics are slowly varying in space.

$$J = -D\nabla\Phi \tag{2.40}$$

Here we have introduced a phenomenological constant of proportionality D. The microscopic definitions of D and v_E are given in the following chapter.

As explained in the section of the Green's function formalism, the simplification of the theory requires considerations of all equations in Fourier space. Therefore the continuity equation Eq. (2.39) and the energy current density Eq. (2.40) are given in Fourier space in the following form

$$-i\Omega\Phi + iQ \cdot J = 1 \tag{2.41}$$

$$J = -iDQ\Phi \tag{2.42}$$

Inserting Eq. (2.40) in Eq. (2.39) we derive the so called diffusion equation

$$\Phi(\Omega, Q) = \frac{i}{\Omega + iQ^2 D(\Omega)} \tag{2.43}$$

which shows the so called diffusion pole structure in the denominator. The explicit discussion of the diffusion pole (see chapter 3) is one of the most interesting parts of the matter and allows to investigate all transport properties of the system even in the presence of dissipation. To understand the influence of diffusion on the energy density relaxation we Fourier transform the diffusion pole from frequency Ω to time t.

$$\Phi(t, Q) \sim e^{-Q^2 D t} \tag{2.44}$$

We introduce the correlation length ξ as defined by the diffusion pole.

$$\frac{1}{\xi^2} = \frac{-i\Omega}{D(\Omega)} \tag{2.45}$$

The correlation length represents the memory of the system in a way that the waves loose their coherence beyond the correlation length. The already discussed Anderson localization of light implies a vanishing diffusion constant $D = 0$. In this case the correlation length becomes the localization length of the energy density Φ, defining the volume within the density is localized in such systems. To express the transport of the energy in the disordered media we need to consider a governing equation for two particle quantities equivalent to the Dyson equation for single particle quantities,

Figure 2.1: *Diagrammatic representation of the Bethe-Salpeter equation Eq. (2.46) for the two particle Green's function $\Phi_{\vec{q}''\vec{q}'}$ in momentum space. $\gamma_{qq''}$ represents the irreducible vertex.*

the Bethe-Salpeter equation, see also Fig. (??). For the two particle Green's function $\Phi_{\vec{q}\vec{q}'}$ the Bethe-Salpeter equation reads

$$\Phi_{\vec{q}\vec{q}'} = G^{\omega}_{q+}G^{*\,\omega-}_{q-}\left(\delta(\vec{q}-\vec{q}') + \int \frac{d^3q''}{(2\pi)^3}\gamma_{qq''}\Phi_{\vec{q}''\vec{q}'}\right) \tag{2.46}$$

introducing the irreducible vertex $\gamma_{qq''}$ which includes all scattering events of the intensity. The irreducible vertex $\gamma_{qq''}$ is the analog to the single particle self-energy Σ in the Dyson equation, which has been discussed above.

The solution of the Bethe-Salpeter equation adopts the scheme presented at the beginning of this section. We construct two independent equations relating energy density and energy density current by utilizing the Bethe-Salpeter equation, the continuity equation and the energy density relation. Combining both we derive the diffusion pole structure of the energy density. A detailed presentation is given in the following chapter. Local energy conservation will be included by means of an exact Ward-Identity as is shown below.

2.4.1 Incoherent and Coherent Contributions

The average two particle Green's function describes the temporal and spatial evolution of the light intensity in a disordered medium. When the two particle Green's function is averaged over a random arrangement of scatterer we can identify three principle contributions: First, the intensity is going from one point to another without any scattering event (*ballistic regime*). Second, the intensity is going from one point to another by a classical process of multiple scattering (*Diffuson*). Third, the intensity is going from one point to another by a coherent process of multiple scattering (*Cooperon*). Naturally this includes interference effects.

Since the considered systems in this thesis rely on multiple scattering processes, ballistic transport will not be discussed further.

Generally spoken, we want to study the occurrence, characteristics and possible disappearance of diffusion (*Anderson localization*). Therefore we have to discuss the remaining two contributions, Diffuson and Cooperon, in order to understand their implication on and relevance to intensity transport in disordered systems.

Ladder Approximation

We must consider the disorder average of the product $\mathbf{G^R G^A}$ of the Greens's functions $\mathbf{G^R}$, $\mathbf{G^A}$, corresponding to all possible multiple scattering sequences such as those represented schematically in Fig. (2.2). There are two consequences of averaging: First we calculate only traveling paths with identical scattering events, meaning the scattering events for two different waves traveling through the media do not have to occur necessarily in the same order. Second, the averaged product $\langle \mathbf{G^R G^A} \rangle$ is replaced by the product of the averaged Green's functions $\langle \mathbf{G^R} \rangle \langle \mathbf{G^A} \rangle$. The distance between two collisions is thus of the order of the elastic mean free path l_s. In the weak disorder regime the collisions are independent and $l_s \gg \lambda$. Consequently the difference in path-lengths between trajectories with non-identical scattering sequences will be at least of order l_s, i.e. much greater than the wavelength λ. The resulting dephasing is very large and these contributions are thus negligible. We therefore retain those contribution of the type represented in Fig. (2.2). This approximation is called the *Diffuson* or *Ladder Approximation*.

2.4. TWO PARTICLE QUANTITIES AND BETHE SALPETER-EQUATION

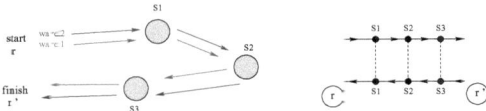

Figure 2.2: *Diffuson or Ladder Approximation.*

Cooperon Approximation

Additionally to the Ladder contributions which we discussed above there is at least one other contribution which has not been taken into account and which has important physical consequences. To see this, consider the product of two Green's functions describing two trajectories which are identical but for the fact that they are covered in exactly opposite directions such as those shown in Fig. (2.3). The phase factors associated with each of the two trajectories are identical, provided the system is time reversal invariant. A process for which the two trajectories are traversed in opposite directions is allowed and thus must contribute equally to the transport.

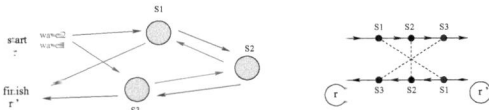

Figure 2.3: *Cooperon contribution*

Because of the characteristic "fan" structure (see Fig. (2.3)) of the diagrams contributing to the *Cooperon*, it is also often called a *maximally crossed Diagram* in the literature. The calculation and disentanglement is shown in Fig. (2.4)).

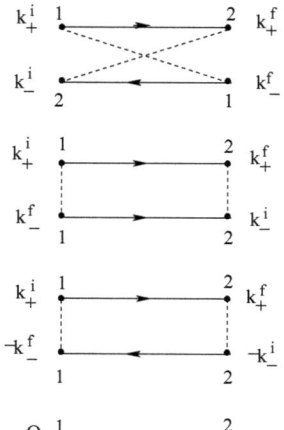

(a) A maximally crossed diagram

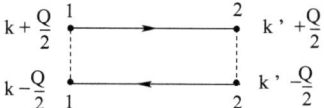

(b) Same as above, but with the bottom line turned around. No effect on evaluation of diagram.

(c) The sign of momenta in the lower line is reversed from the above. This leaves the diagram the same if the system has time reversal symmetry.

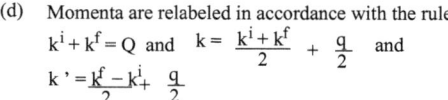

(d) Momenta are relabeled in accordance with the rule $k^i + k^f = Q$ and $k = \frac{k^i + k^f}{2} + \frac{q}{2}$ and $k' = \frac{k^f - k^i}{2} + \frac{q}{2}$

Figure 2.4: *To evaluate a maximally crossed diagram (a), we flip the lower time axis which has no effect on the evaluation of the diagram (b). The sign of the momenta in the lower (-) channel is reversed (c), provided time reversal invariance. Now the momenta are relabeled (d) as shown in the figure.*

Chapter 3

Light Transport in Infinite Three Dimensional Disordered Media with Absorption or Amplification

The first goal of this thesis is to gain a theory for scalar wave propagation and light intensity transport in tree dimensional random dielectric media with gain and/or absorption. Special emphasis is put on the detailed presentation of the solution method of the governing Bethe-Salpeter equation. The proximity of the Anderson transition is treated by a self-consistent Cooperon resummation, representing the repeated self-interference, based on the theory of Vollhardt and Wölfle [60]. In this modified form it accounts for dissipation, as well as energy conservation in terms of a generalized Ward identity [61]. The incorporated approximations are discussed in detail with respect to the range of applicability, especially regarding optical gain. As there is no full description in literature for such a theoretical model whatsoever, this chapter provides a lot of technical details which shall lead the interested reader through the calculations.

3.1 Introduction

Light propagation and light intensity transport in random media remain an active field of research in theory [17, 18, 19, 20] and experiment [21, 22, 23]. In recent years special interest has arisen on the effects of absorption and on lasing in disordered [24, 25, 27, 28, 29, 30, 31] and periodic [32, 33] structures. On the theoretical side, many authors [18, 34, 35], including also some of our previous work [19], employ methods which are commonly used in calculations known from electronic systems in solid state physics. Examples are single-particle Green's functions [82, 119] as well as higher-order Green's functions [82, 119, 18] and their respective equations of motion, e.g. the Bethe-Salpeter equation, including their solution ansatzes. While those electronic systems are well documented in the literature, see e.g. [36], the photonic counterparts, however, experience a lack of coherent and detailed presentation. The purpose of this chapter is to provide a detailed presentation of the theory for photonic systems to be used by the community.

There are subtle but important differences between electronic and photonic systems, such as the equation of motion for photons being second order in time as compared to the first order in time Schrödinger equation of electrons. This leads to a frequency dependent random potential for photons, as will be shown in section 3.2 and further renormalizes the transport velocity of the light intensity as well as the form of the diffusion constant in section 3.5. Moreover, non-conservation of particles is unknown in electronic systems in contrast to the emission and absorption of photons.

The presented work in this chapter is organized as follows. First, we derive in detail how stimulated emission, described by laser rate equations, enters into the transport theory as an imaginary part of the dielectric function. In a second step we continue with a detailed documentation of the solution of the Bethe-Salpeter equation for photonic correlators, which, to the best of our knowledge, has not been published before. Differences with respect to the electronic case will be pointed out all along this course where appropriate.

Special emphasis has to be put on investigating the role that Anderson localization takes in disordered emitting media. Anderson or strong localization [37] has been shown [59, 39] to arise from repeated self-interference of diffusive modes. In experiments [40, 41, 42] this may have been tested. Since intensity diffusion is solely based on energy conservation, a rigorous and consistent framework is needed to describe the interplay between coherent amplification/absorption and localization. Especially since coherent amplification is expected to enhance transmission whereas localization tends to stop light transport altogether. This interesting subject [43] is also discussed in the context of random lasing [28, 44], where it has been shown by measuring the photon statistics [45] that the laser emission is due to coherent feedback and occurs at spatially confined spots in the sample. A theoretical attempt [46] to independently explain such phenomena has been proposed based on scatterers statistically forming ring resonators within the sample, which may provide a feedback mechanism.

3.2 Light Matter Interaction

In general, the propagation of electromagnetic waves in dielectric media may be described with the help of Maxwell's equations, which read again (in Gaussian units)

$$\begin{aligned} \nabla \cdot D &= 4\pi\rho & \nabla \times E + \frac{1}{c}\frac{\partial B}{\partial t} &= 0 \\ \nabla \cdot B &= 0 & \nabla \times H &= \frac{4\pi}{c}j + \frac{1}{c}\frac{\partial D}{\partial t} \end{aligned} \quad (3.1)$$

with the electric displacement D and magnetic induction B given by

$$D = E - 4\pi P \qquad H = B - 4\pi M \quad (3.2)$$

where P is the medium's polarization and M its magnetization. Combining Maxwell's equations, Eq. (3.1), one obtains the wave equation as discussed in the previous chapter

$$\nabla^2 E(\vec{r},t) - \frac{1}{c^2}\frac{\partial^2}{\partial t^2}D(\vec{r},t) = \frac{4\pi}{c}\frac{\partial}{\partial t}j(\vec{r},t) \quad (3.3)$$

describing the propagating electric field in the presence of dielectrics, where the absence of free charges and currents has been assumed as well as a non-magnetic medium. In the following we discuss how wave equation we derived here, Eq. (3.3), is modified by stimulated light emission due to the presence of a laser active medium.

Within a semi-classical approach a lasing system may be described by the so-called Maxwell-Bloch equations (MBE), treating population inversion, polarization and light propagation as a coupled set of equations. In absence of quantum noise, the MBE for 4-level atoms within rotating-wave and the slowly varying envelope approximations can be written as [47]

$$\frac{\partial W}{\partial t}(\vec{r},t) = \gamma_\parallel \left[\Gamma_p - W(\vec{r},t)\right] \quad (3.4)$$

$$+ \frac{i}{4\hbar}\left[P^*(\vec{r},t)E(\vec{r},t) - P(\vec{r},t)E^*(\vec{r},t)\right]$$

$$\frac{\partial P}{\partial t}(\vec{r},t) = -\gamma_\perp P(\vec{r},t) + W(\vec{r},t) \cdot \frac{i\mu^2}{\hbar}E(\vec{r},t) \quad (3.5)$$

$$\frac{\partial^2 P}{\partial t^2}(\vec{r},t)e^{-i\omega t} = \frac{1}{4\pi}\left[c^2\nabla^2 - \epsilon\frac{\partial^2}{\partial t^2} - \eta\frac{\partial}{\partial t}\right]E(\vec{r},t)e^{-i\omega t} \quad (3.6)$$

where the dynamical variables are the macroscopic electric field $E(\vec{r},t)$, the population inversion density $W(\vec{r},t)$ and the polarization $P(\vec{r},t)$. In Eq. (3.4), the population inversion is achieved by external pumping Γ_p and it is decaying (spontaneously) at the rate $\gamma_\parallel$. The polarization in Eq. (3.5) is generated by the atomic transition dipole moment μ. It decays at a (collisional dephasing) rate $\gamma_\perp$ and drives the macroscopic electric field $E(\vec{r},t)$ in Eq. (3.6), which may experience dissipation

3.2. LIGHT MATTER INTERACTION

by intrinsic absorption or loss out of the feedback mechanism. Both effects are combined here in the dissipation term η.

Concerning the laser dynamics, we observe that Eq. (3.5) describes a crucial part of the polarization dynamics on which we want to concentrate now. In general, there are no specific restrictions to, e.g. the various decay rates of a lasing system. Such unrestricted lasers are typically classified as class C-type laser [48]. However, existing lasers often display certain restrictions on the involved time scales, which may be used to further classify or categorize a given laser. One particular interesting case is when $\gamma_\perp$ is the dominating time scale, a so-called class B laser [48]. In this case the polarization of the medium relaxes and reaches saturation so quickly that the polarization may be assumed to exhibit no dynamics of its own but simply following E and W in an adiabatic fashion. This approximation implies

$$|\gamma_\perp P(\vec{r},t)| \gg \left|\frac{\partial P(\vec{r},t)}{\partial t}\right| \qquad (3.7)$$

so that the above equation. Eq. (3.5), now reads

$$0 = -\gamma_\perp P(\vec{r},t) + i\kappa W(\vec{r},t) \cdot E(\vec{r},t), \qquad (3.8)$$

where we introduced $\kappa = \frac{t^2}{4}$, so that the polarization is eventually found to be

$$P(\vec{r},t) = \frac{i\kappa}{\gamma_\perp} W(\vec{r},t) \cdot E(\vec{r},t). \qquad (3.9)$$

The above equation, Eq. (3.9), may now be used together with Eq. (3.2) to derive an expression for the electric displacement D

$$D(\vec{r},t) = \left(1 - i\frac{4\pi\kappa}{\gamma_\perp}W(\vec{r},t)\right)E(\vec{r},t) \stackrel{\text{def.}}{=} \epsilon(\vec{r})E(\vec{r},t) \qquad (3.10)$$

which serves also as a definition for the dielectric function ϵ. It is therefore given by the important relation

$$\epsilon(\vec{r}) = 1 - i\frac{4\pi\kappa}{\gamma_\perp}W(\vec{r},t). \qquad (3.11)$$

Bearing in mind that the population inversion density may be expanded in the electric field strength according to

$$W(E(\vec{r},t)) = W_0 - W_1 |E(\vec{r},t)|^2 + \ldots \qquad (3.12)$$

the dielectric function decomposes into

$$\begin{aligned} \epsilon(\vec{r}) &= 1 - i\frac{4\pi\kappa}{\gamma_\perp}W_0 + i\frac{4\pi\kappa}{\gamma_\perp}W_1 |E(\vec{r},t)|^2 \\ &= \epsilon_L + \epsilon_{NL}\left[E(\vec{r},t)\right] \equiv \epsilon\left[E(\vec{r},t)\right] \end{aligned} \qquad (3.13)$$

where the now complex-valued function ϵ_L describes the linear optics regime in the presence of gain, whereas $\epsilon_{NL}[E(\vec{r},t)]$ incorporates non-linear optical effects, which are to lowest order quadratic in the electric field. With this decomposition of the dielectric function, we find

$$\begin{aligned} \frac{\partial^2}{\partial t^2}D(\vec{r},t) &= \frac{\partial^2}{\partial t^2}\left(\epsilon\left[E(\vec{r},t)\right]E(\vec{r},t)\right) \\ &= \epsilon_L \frac{\partial^2}{\partial t^2}E(\vec{r},t) + \frac{\partial^2}{\partial t^2}\left(\epsilon_{NL}\left[E(\vec{r},t)\right]E(\vec{r},t)\right), \end{aligned} \qquad (3.14)$$

which implies the following scenario: While the light intensity, i.e. also the macroscopic electric field $E(\vec{r},t)$, is exponentially built up within the class B lasing system due to the external pumping Γ_p, there is a regime in which the light intensity is still small enough to prevent the system from non-linear contributions to the dielectric function, meaning that up to this point of validity of linear optics the system is correctly and rigorously described by a dielectric function which may as well be parameterized as

$$\epsilon = \text{Re}\,\epsilon - i\text{Im}\,\epsilon \qquad (3.15)$$

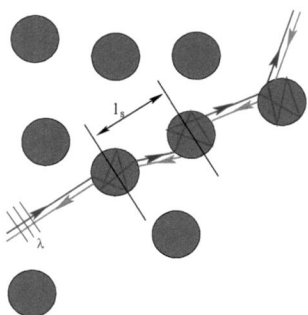

Figure 3.1: *Model of randomly distributed scatterers in homogeneous background.*

where Im ϵ is uniquely determined by the population inversion as discussed in Eq. (3.13). In this way the parameterization, Eq. (3.15), is restricted by the onset on non-linear behavior in the response, as expected. Within this range of validity, the change to the wave equation for the electric field, Eq. (3.3), consists in replacing the dielectric displacement D by $D = \epsilon E$ according to Eq. (3.14) and Eq. (3.15), eventually yielding

$$\nabla^2 E(\vec{r},t) - \frac{\epsilon(\vec{r})}{c^2}\frac{\partial^2}{\partial t^2}E(\vec{r},t) = \frac{4\pi}{c}\frac{\partial}{\partial t}j(\vec{r},t). \qquad (3.16)$$

In the following we want to restrict ourselves to this regime of a class B laser system. An analogous reasoning applies also for (linear) absorption, resulting in the opposite sign of the imaginary part of Eq. (3.15). Therefore we discuss in the remainder of this article the propagation of light described by Eq. (3.16) including both absorption/emission depending on the sign of the imaginary part of the dielectric function.

3.3 Setup and Model

Systems of significant experimental relevance [22, 23, 27, 28, 29, 30, 31] consist of (almost) spherical scatterers embedded into a background medium forming some emulsion. For a theoretical description we therefore consider identical spherical scatterers located at random positions (see Fig. 3.1). The scatterers as well as the background medium are respectively assumed to be homogeneous themselves and hence will be described by dielectric constants ϵ_s and ϵ_b, respectively. Within the above discussed range of validity for photon emission (of course also for photon absorption) the dielectric function may be assumed to have a finite positive or negative imaginary part, so that we have in general Im $\epsilon_s \neq 0 \neq$ Im ϵ_b, as outlined in the previous section. Throughout this thesis we neglect polarization effects of light and therefore consider the *scalar* wave equation which here has been Fourier transformed from time t to light frequency ω and reads for the scalar field $\Psi_\omega(\vec{r})$, as it is obvious from Eq. (3.16),

$$\frac{\omega^2}{c^2}\epsilon(\vec{r})\Psi_\omega(\vec{r}) + \nabla^2\Psi_\omega(\vec{r}) = -i\omega\frac{4\pi}{c^2}j_\omega(\vec{r}) , \qquad (3.17)$$

where c denotes again the vacuum speed of light and $j_\omega(\vec{r})$ an external source. The dielectric constant $\epsilon(\vec{r}) = \epsilon_b + \Delta\epsilon V(\vec{r})$, where the dielectric contrast has been defined as $\Delta\epsilon = \epsilon_s - \epsilon_b$, describes the arrangement of scatterers through the function $V(\vec{r}) = \sum_{\vec{R}} S_{\vec{R}}(\vec{r})$, with $S_{\vec{R}}(\vec{r})$ a localized shape function at random locations $\vec{R}$. The intensity is then related to the field-field-correlation function, often referred to as the four-point-correlation, $\langle \Psi(\vec{r}, t)\Psi^*(\vec{r}', t')\rangle$. To calculate the field-field-correlation the Green's function formalism is best suited, the (single-particle) Green's function is related to the (scalar) electrical field by

$$\Psi(\vec{r}, t) = \int d^3 r' \int dt' \, G(\vec{r}, \vec{r}'; t, t') j(\vec{r}', t'). \tag{3.18}$$

The Fourier transform of the retarded, disorder averaged single-particle Green's function of Eq. (3.17) reads,

$$G^\omega_{\vec{q}} = \frac{1}{\epsilon_b(\omega/c)^2 - |\vec{q}|^2 - \Sigma^\omega_{\vec{q}}}, \tag{3.19}$$

where the retarded self-energy $\Sigma^\omega_{\vec{q}}$ arises from scattering off the random "potential" $-(\omega/c)^2(\epsilon_s - \epsilon_b)V(\vec{r})$, which is obviously frequency dependent. Using Green's functions, the mode density $N(\omega)$ may be expressed as $N(\omega) = -(\omega/\pi) \operatorname{Im} G^\omega_0$, with the abbreviation used throughout this publication $G^\omega_0 \equiv \int d^3q/(2\pi)^3 \, G^\omega_{\vec{q}}$.

In order to study the transport of the above introduced field-field-correlation we consider the 4-point correlation function, defined in terms of the non-averaged Green's functions $\hat{G}$, $\hat{G}^*$ in momentum and frequency space as $\Phi^{\omega}_{\vec{q}\vec{Q}}(\vec{Q}, \Omega) = \langle \hat{G}^{\omega_+}_{\vec{q}_+\vec{q}'_+} \hat{G}^{\omega_-\,*}_{\vec{q}'_-\vec{q}_-}\rangle$. Here we have introduced the usual [61] center-of-mass $(\vec{q}, \omega)$ and relative $(\vec{Q}, \Omega)$ frequencies and momenta: The variables Ω, $\vec{Q}$ are associated with the time and position dependence of the averaged energy density, with $\hat{Q} = \vec{Q}/|\vec{Q}|$, while $\omega_\pm = \omega \pm \Omega/2$ and $\vec{q}_\pm = \vec{q} \pm \vec{Q}/2$ etc. are the frequencies and momenta of in- and out-going waves, respectively.

3.4 Light Propagation in Ladder Approximation

Here we review the so-called ladder-approximation for both the photonic particle-hole vertex, as depicted in Fig. 3.2, and the 4-point correlation function, as needed in section 3.5.2.

To describe diffusive processes, explicit results for the two-particle Green's function and the respective vertex also require the expansion of the single-particle Green's function for small arguments, being either or both $|\vec{Q}|$ and Ω. This limit represents the long time and long distance regime, where diffusive behavior shows up. The explicit dependence of the retarded and advanced Green's function on the relevant, diffusive, variables $\vec{Q}$ and Ω is given by

$$G^R_{p_+}(\omega_+) = \frac{1}{\epsilon_b^R(\omega + \frac{\Omega}{2})^2 - (\vec{p} + \frac{\vec{Q}}{2})^2 - \Sigma^R} \tag{3.20}$$

$$G^A_{p_-}(\omega_-) = \frac{1}{\epsilon_b^A(\omega - \frac{\Omega}{2})^2 - (\vec{p} - \frac{\vec{Q}}{2})^2 - \Sigma^A}, \tag{3.21}$$

which is to be distinguished from the electronic Green's function by its frequency dependence in ω and through $\Sigma^{R/A}$. This difference traces back to corresponding governing equation, being either the electronic Schrödinger equation or the bosonic Klein-Gordon equation. Employing now the identity

$$G^R_{p_+}(\omega_+) G^A_{p_-}(\omega_-) = \frac{G^A_{p_-}(\omega_-) - G^R_{p_+}(\omega_+)}{[G^R_{p_+}(\omega_+)]^{-1} - [G^A_{p_-}(\omega_-)]^{-1}}, \tag{3.22}$$

the product of the retarded and advanced single-particle Green's function $G^R_{p_+}(\omega_+) G^A_{p_-}(\omega_-)$ may therefore be expanded for small $\vec{Q}$ and Ω to yield

$$G^R_{p_+}(\omega_+)G^A_{p_-}(\omega_-) \xrightarrow[(0,0)]{(\vec{Q},\Omega)} \frac{\mathrm{Im}\, G_{\vec{p}}}{\mathrm{Im}\,\Sigma - \omega^2 \mathrm{Im}\,\epsilon_b} \qquad (3.23)$$

$$+ \ \Omega\,\mathrm{Re}\,(\epsilon_b)\,\frac{i\omega\,\mathrm{Im}\,G_{\vec{p}}}{(\mathrm{Im}\,\Sigma - \omega^2 \mathrm{Im}\,\epsilon_b)^2}$$

$$- \ Q\left[\vec{p}\cdot\hat{Q}\right]\frac{i\,\mathrm{Im}\,G_{\vec{p}}}{(\mathrm{Im}\,\Sigma - \omega^2 \mathrm{Im}\,\epsilon_b)^2}$$

$$- \ Q^2\,2\left(\vec{p}\cdot\hat{Q}\right)^2\frac{\mathrm{Im}\,G^2_{\vec{p}}}{(\mathrm{Im}\,\Sigma - \omega^2 \mathrm{Im}\,\epsilon_b)}.$$

When the momentum dependence of the Green's function is integrated out, the third term of the right hand side of the above equation, Eq. (3.23), will not survive due to its odd symmetry, eventually yielding

$$\int\frac{\mathrm{d}^3p}{(2\pi)^3}G^R_{p_+}(\omega_+)G^A_{p_-}(\omega_-) = \frac{\mathrm{Im}\,G_0}{\mathrm{Im}\,\Sigma - \omega^2 \mathrm{Im}\,\epsilon_b} \qquad (3.24)$$

$$+ \ \Omega\,\mathrm{Re}\,(\epsilon_b)\,\frac{i\omega\,\mathrm{Im}\,G_0}{(\mathrm{Im}\,\Sigma - \omega^2 \mathrm{Im}\,\epsilon_b)^2}$$

$$- \ Q^2\,\frac{2\int\frac{\mathrm{d}^3p}{(2\pi)^3}(\vec{p}\cdot\hat{Q})^2\mathrm{Im}\,G^2_{\vec{p}}}{(\mathrm{Im}\,\Sigma - \omega^2 \mathrm{Im}\,\epsilon_b)^2},$$

where G_0 is defined to be $G_0 \equiv \int\frac{\mathrm{d}^3p}{(2\pi)^3}G_{\vec{p}}$.

After having discussed the basic expansions, let us now turn to the discussion of the two-particle vertex in the ladder approximation as illustrated in Fig. 3.2. The shown exponential series can be summed up to yield the expression

$$\Gamma_L(\Omega,\vec{Q}) = \frac{\gamma_0}{1 - \gamma_0\int\frac{\mathrm{d}^3q}{(2\pi)^3}G^R_{p_+}(\omega_+)G^A_{p_-}(\omega_-)}, \qquad (3.25)$$

where γ_0 is the bare irreducible interaction vertex. Expanding the product of the two single-particle Green's functions in the denominator of Eq. (3.25) up to the first non-vanishing order in momentum $\vec{Q}$ and frequency Ω according to Eq. (3.23), the following expression is obtained

$$\Gamma_L(\Omega,\vec{Q}) = \left(\frac{v_E c_p}{\omega}\right)\frac{\gamma_0(\Delta\Sigma - \omega^2\Delta\epsilon)/\omega}{\Omega + iQ^2 D_0 + iD_0 x_a^{-2}}, \qquad (3.26)$$

where the coefficient D_0 of the squared momentum, the so-called bare diffusion constant, describes the diffusive relaxation of a given exited mode and reads

$$D_0 = \frac{2v_E c_p}{\pi N(\omega)}\int\frac{\mathrm{d}^3q}{(2\pi)^3}\,[\hat{q}\cdot\hat{Q}]^2(\mathrm{Im}\,G^\omega_{\vec{q}})^2. \qquad (3.27)$$

The expression for the vertex $\Gamma_L(\Omega,\vec{Q})$ in Eq. (3.26) clearly shows the diffusion pole structure except for the last term in the denominator. This term, $D_0 x_a^{-2}$, represents the so-called mass-term, accounting for, non-diffusive, losses out of a given diffusive Q-mode. Such losses are due to the dissipative (absorbing) properties of the medium itself, in which the diffusion takes place. This mass-term is explicitely given by

$$x_a^{-2} = \frac{r_\epsilon A_\epsilon - 2\omega^2\,\mathrm{Im}\,\epsilon_b}{2\,\mathrm{Re}\,\epsilon_b - A_\epsilon B_\epsilon/\omega}\frac{1}{\omega D_0}. \qquad (3.28)$$

The prefactor of the actual diffusion pole in Eq. (3.26) contains the phase velocity c_p of the disordered medium

$$c_p = \mathrm{Re}\,\frac{c}{\sqrt{\epsilon_b - \Sigma^\omega_0\frac{c^2}{\omega^2}}} \qquad (3.29)$$

3.4. LIGHT PROPAGATION IN LADDER APPROXIMATION

as well as the energy transport velocity v_E given as

$$v_E = \frac{c^2}{c_p \operatorname{Re} \epsilon_b} \frac{1}{1 + \Delta(\omega)} \tag{3.30}$$

and describing the speed at which the energy contained in the wave field is expected to pass through the medium. The correction or renormalization $\Delta(\omega)$ reads

$$\Delta(\omega) = B_\epsilon A_\epsilon + i r_\epsilon \partial_\Omega A_\epsilon(\Omega) \tag{3.31}$$

with the following abbreviations introduced

$$u_\epsilon = \frac{\operatorname{Im}(\Delta \epsilon \Sigma^\omega)}{\operatorname{Im}(\Delta \epsilon G_0^\omega)}, \qquad r_\epsilon = \operatorname{Im} \Delta \epsilon / \operatorname{Re} \Delta \epsilon, \tag{3.32}$$

$$A_\epsilon = 2[u_\epsilon \operatorname{Re} G_o + \operatorname{Re} \Sigma_o] \qquad B_\epsilon = \frac{(\operatorname{Re} \Delta \epsilon)^2 + (\operatorname{Im} \Delta \epsilon)^2}{2 \omega^2 (\operatorname{Re} \Delta \epsilon)^2}.$$

Furthermore, for later use and to keep shorthand definitions in one place, we state here the abbreviations

$$\tilde{A} = \frac{c_p v_E}{\omega} \left[1 - \frac{\omega^2 \operatorname{Im} \epsilon_b}{u_\epsilon \operatorname{Im} G_0} + \frac{r_\epsilon A_\epsilon}{u_\epsilon \operatorname{Im} G_0} \right] \tag{3.33}$$

$$\Lambda(\omega) = i \omega^2 \operatorname{Im} \epsilon_b - i r_\epsilon A_\epsilon$$

$$g_\omega^{(0)} = \frac{2\omega}{c^2} \operatorname{Im} \epsilon_b \qquad g_\omega^{(1)} = \frac{4\omega}{c^2} \operatorname{Re} \epsilon_b.$$

In the remaining part of this section we need to discuss the evaluation of the energy density correlation $P_E^\omega(\vec{Q}, \Omega)$, the current-density-correlation $J_E^\omega(\vec{Q}, \Omega)$, and the two-particle Green's function, respectively, within the ladder approximation.
Turning first to the energy density correlation $P_E^\omega(\vec{Q}, \Omega)$ as defined in Eq. (3.45). Within the ladder approximation according to the illustration in Fig. 3.2, it is given by

$$P_E^{\omega \, L}(\vec{Q}, \Omega) = \left(\frac{\omega}{c_p}\right)^2 \int \frac{d^3 q}{(2\pi)^3} \left[G_{\vec{q}_+}(\vec{Q}, \Omega) G_{\vec{q}_-}^*(\vec{Q}, \Omega) \right]^2 \Gamma_L$$

$$= \left(\frac{\omega}{c_p}\right)^2 \frac{1}{\tilde{\gamma}_0^2} \Gamma_L, \tag{3.34}$$

where the superscript L indicates that the correlation function is meant to be approximated by the ladder diagrams only. In the last step of Eq. (3.34) we have used the expansion of the single-particle Green's function for small momenta $\vec{Q}$ as given by Eq. (3.23) and computed $P_E^{\omega \, L}(\vec{Q}, \Omega)$ up to the same order of the expansion as we did for Γ_L. Furthermore, in Eq. (3.34) we introduced the renormalized vertex $\tilde{\gamma}_0$ given by

$$\tilde{\gamma}_0 = \gamma_0 + f_\omega(\Omega) \frac{(\operatorname{Re} \gamma_0 G_0 + \operatorname{Re} \Sigma)}{\operatorname{Im} G_0} - \frac{\omega^2 \operatorname{Im} \epsilon_b}{\operatorname{Im} G_0} \tag{3.35}$$

where γ_0 is the bare vertex, whereas $f_\omega(\Omega)$ arises from the Ward identity and will be defined in Eq. (3.43).
The current-density-correlation $J_E^\omega(\vec{Q}, \Omega)$, which will be defined in Eq. (3.46), yields, when evaluated within the ladder approximation,

$$J_E^{\omega \, L}(\vec{Q}, \Omega) = \left(\frac{\omega v_E}{c_p}\right) \int \frac{d^3 q}{(2\pi)^3} (\vec{q} \cdot \hat{Q}) G_{\vec{q}_+} G_{\vec{q}_-}^* \int \frac{c^3 q'}{(2\pi)^3} G_{\vec{q}'_+} G_{\vec{q}'_-}^* \Gamma_L. \tag{3.36}$$

Following the above strategy and expanding the product $G_{\vec{q}'_+} G_{\vec{q}'_-}^*$ under the second integral up to first order in $\vec{q}'$ as shown in Eq. (3.23) one obtains the expression

$$J_E^{\omega \, L}(\vec{Q}, \Omega) = \left(\frac{\omega v_E}{c_p}\right) \frac{1}{\tilde{\gamma}_0} \Gamma_L \int \frac{d^3 q}{(2\pi)^3} (\vec{q} \cdot \hat{Q}) G_{\vec{q}_+} G_{\vec{q}_-}^*. \tag{3.37}$$

Employing now the same expansion to the remaining product of the Green's function one eventually finds

$$J_E^{\omega\,L}(\vec{Q},\Omega) = \left(\frac{\omega v_E}{c_p}\right)\frac{\Gamma_L}{\tilde{\gamma}_0}\int\frac{d^3q}{(2\pi)^3}(\vec{q}\cdot\hat{Q})\frac{1}{2}\frac{\Delta G_{\vec{q}}^2(\vec{q}\cdot\hat{Q})Q}{\tilde{\gamma}_0\Delta G_0}, \quad (3.38)$$

where the abbreviation $\Delta G \equiv G - G^*$ has been introduced and will be used throughout this chapter. Turning now towards the 4-point correlation function, according to Fig. 3.2 in ladder approximation it is given as

$$\Phi_{\vec{q}}^L = \int\frac{d^3q'}{(2\pi)^3}\,\Phi_{\vec{q}\vec{q}'}^L = \left[G_{\vec{q}_+}G_{\vec{q}_-}^*\right]\Gamma_L\left[\int\frac{d^3q'}{(2\pi)^3}G_{\vec{q}'_+}G_{\vec{q}'_-}^*\right] \quad (3.39)$$

Employing again the momentum expansion of the single-particle Green's function the above equation, Eq. (3.39) may be simplified to yield the final expression

$$\Phi_{\vec{q}}^L = \frac{\Delta G_{\vec{q}}}{\tilde{\gamma}_0^2\Delta G_0}\Gamma_L + \frac{1}{2}\frac{\Delta G_{\vec{q}}^2(\vec{q}\cdot\hat{Q})Q}{\tilde{\gamma}_0^2\Delta G_0}\Gamma_L. \quad (3.40)$$

3.5 Theory of Transport and Localization

The intensity correlation, or disorder averaged particle-hole Green's function, $\Phi_{\vec{q}\vec{q}'}^\omega(\vec{Q},\Omega)$ obeys the so-called Bethe-Salpeter equation

$$\Phi_{\vec{q}\vec{q}'}^\omega = G_{q_+}^R(\omega_+)G_{q_-}^A(\omega_-)\left[1 + \int\frac{d^3q''}{(2\pi)^3}\,\gamma_{qq''}\,\Phi_{\vec{q}''\,\vec{q}'}^\omega\right], \quad (3.41)$$

where we have suppressed the dependence on the variables $(\vec{Q},\Omega)$ in order to maintain a clearer notation, so $\Phi_{\vec{q}\vec{q}'}^\omega$ reads $\Phi_{\vec{q}\vec{q}'}^\omega(\vec{Q},\Omega)$ etc. By utilizing the known averaged single particle Green's function, c.f. Eq. (3.19), on the left-hand side of Eq. (3.41) the Bethe-Salpeter equation may be rewritten as kinetic equation,

$$\left[\omega\Omega\frac{\text{Re}\epsilon_b}{c^2} - Q(\vec{q}\cdot\hat{Q}) + \frac{i}{c^2\tau^2}\right]\Phi_{\vec{q}\vec{q}'}^\omega =$$
$$-i\text{Im}G_{\vec{q}}^\omega\left[1 + \int\frac{d^3q''}{(2\pi)^3}\,\gamma_{\vec{q}\vec{q}''}^\omega\Phi_{\vec{q}'\vec{q}}^\omega\right]. \quad (3.42)$$

In order to analyze the correlation function's long-time ($\Omega \to 0$) and long-distance ($|\vec{Q}| \to 0$) behavior, terms of order $O(\Omega^2, Q^3, \Omega Q)$ have been neglected here and throughout this chapter. Eq. (3.42) contains both, the *total* quadratic momentum relaxation rate $1/\tau^2 = c^2\,\text{Im}(\epsilon_b\omega^2/c^2 - \Sigma^\omega)$ (due to absorption in the background medium as well as impurity scattering) and the irreducible two-particle vertex function $\gamma_{\vec{q}\vec{q}'}^\omega(\vec{Q},\Omega)$. To solve this equation, the technique of expansion into moments is used. The technical details of this expansion are discussed in the following subsection 3.5.1

Furthermore it is to be noted that the energy conservation is implemented into the solution of the Bethe-Salpeter equation in a field theoretical sense by a Ward identity (WI) which has been derived for the photonic case in Ref. [61], and which for scalar waves takes the exact form

$$\Sigma_{\vec{q}_+}^{\omega_+} - \Sigma_{\vec{q}_-}^{\omega_-\,*} - \int\frac{d^3q'}{(2\pi)^3}\left[G_{\vec{q}'_+}^{\omega_+} - G_{\vec{q}'_-}^{\omega_-\,*}\right]\gamma_{\vec{q}'\vec{q}}^\omega(\vec{Q},\Omega) \quad (3.43)$$
$$= f_\omega(\Omega)\left[\text{Re}\Sigma_{\vec{q}}^\omega + \int\frac{d^3q'}{(2\pi)^3}\,\text{Re}G_{\vec{q}'}^\omega\,\gamma_{\vec{q}'\vec{q}}^\omega(\vec{Q},\Omega)\right].$$

The right-hand side of Eq. (3.43) represents reactive effects (real parts), originating from the explicit ω^2-dependence of the photonic random "potential". In conserving media ($\text{Im}\epsilon_b = \text{Im}\epsilon_s = 0$) these terms renormalize the energy transport velocity v_E relative to the average phase velocity c_p without destroying the diffusive long-time behavior[35, 61]. In presence of loss or gain, however, these effects are enhanced via the prefactor $f_\omega(\Omega) = (\omega\Omega\text{Re}\Delta\epsilon + i\omega^2\text{Im}\Delta\epsilon)/(\omega^2\text{Re}\Delta\epsilon + i\omega\Omega\text{Im}\Delta\epsilon)$, which now does not vanish in the limit $\Omega \to 0$.

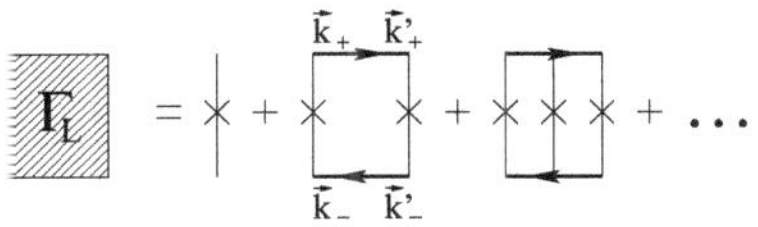

3.5. THEORY OF TRANSPORT AND LOCALIZATION

Figure 3.2: *Ladder approximation of the total particle-hole vertex. The diagrams on the left-hand side form a geometrical series and may therefore easily be summed up analytically.*

3.5.1 Expansion of Two-particle Green's Function into Moments

In order to extract a diffusion pole structure out of the Bethe-Salpeter equation, Eq. (3.41), the correlator or equivalently the $\vec{q}\,'$ integrated correlator

$$\Phi^\omega_{\vec{q}} = \int \frac{d^3 q'}{(2\pi)^3} \Phi^\omega_{\vec{q}\vec{q}\,'} \tag{3.44}$$

has to be decoupled from the momentum dependent pre factors with the help of some approximation scheme. In this subsection we discuss this procedure in analogy to the argumentation for electronic correlations presented in reference [36].
Such an approximation must obey the results of the so-called ladder approximation (describing simple diffusion) as well as it must incorporate the set of physical relevant variables involved in observed phenomena.
In a first step let us define the energy density correlation $P^\omega_E(\vec{Q}, \Omega)$ and the current-density-correlation $J^\omega_E(\vec{Q}, \Omega)$, which involve the first and second moment of the correlation function $\Phi^\omega_{\vec{q}}$ from Eq. (3.44), respectively,

$$P^\omega_E(\vec{Q}, \Omega) = \left(\frac{\omega}{c_p}\right)^2 \int \frac{d^3 q}{(2\pi)^3} \int \frac{d^3 q'}{(2\pi)^3} \Phi^\omega_{\vec{q}\vec{q}\,'} \tag{3.45}$$

$$J^\omega_E(\vec{Q}, \Omega) = \left(\frac{\omega v_E}{c_p}\right) \int \frac{d^3 q}{(2\pi)^3} \int \frac{d^3 q'}{(2\pi)^3} (\vec{q} \cdot \hat{Q}) \Phi^\omega_{\vec{q}\vec{q}\,'}. \tag{3.46}$$

The projection of the correlator $\Phi^\omega_{\vec{q}}$, Eq. (3.44), onto the moments $P^\omega_E(\vec{Q}, \Omega)$ as defined in Eq. (3.45), and $J^\omega_E(\vec{Q}, \Omega)$, shown in Eq. (3.46), is therefore given by

$$\int \frac{d^3 q'}{(2\pi)^3} \Phi^\omega_{\vec{q}\vec{q}\,'} = \frac{A(\vec{q})}{c_1 \int \frac{d^3 q'}{(2\pi)^3} A(\vec{q}\,')} P^\omega_E(\vec{Q}, \Omega) \tag{3.47}$$

$$+ \frac{B(\vec{q})(\vec{q} \cdot \hat{Q})}{c_2 \int \frac{d^3 p'}{(2\pi)^3} B(\vec{q}\,')(\vec{q}\,' \cdot \hat{Q})^2} J^\omega_E(\vec{Q}, \Omega),$$

where the projection coefficients $A(\vec{q})$ and $B(\vec{q})$ are to be determined in the following together with the normalization constants c_1 and c_2. Since those expansion coefficients $A(\vec{q})$ and $B(\vec{q})$ in Eq. (3.47) should behave uncritically under localization, they may be determined using the simple ladder approximation, where all expressions are known exactly.
We reviewed the so-called ladder-approximation for both the photonic particle-hole vertex, as depicted in Fig. 3.2, and the 4-point correlation function of section 3.4, since there are severe differences with respect to the electronic system.

3.5.2 Computing the Coefficients of the Expansion into Moments

After evaluating the relevant quantities within the ladder approximation in section 3.4, we are now in a position to calculate the coefficients of the proposed expansion of the two-particle Green's function in Eq. (3.47). As already pointed out above, the desired expansion has to be valid independently of a particularly chosen approximation, as the ladder approximation is. Therefore we take advantage of the ladder approximation where all quantities are known exactly and use this to compute the still unknown coefficients $A(\vec{q})$ and $B(\vec{q})$. In particular combining the expressions for energy density correlation $P_E^\omega(\vec{Q},\Omega)$ in Eq. (3.34) and the current-density-correlation $J_E^\omega(\vec{Q},\Omega)$ in Eq. (3.38) with the proposed expansion of the two-particle Green's function in Eq. (3.40) we obtain the relation, valid within the ladder approximation

$$\int \frac{\mathrm{d}^3 q'}{(2\pi)^3} \Phi_{\vec{q}\vec{q}'} = \frac{A(\vec{q})}{c_1 \int \frac{\mathrm{d}^3 q'}{(2\pi)^3} A(\vec{q}')} \left(\frac{\omega}{c_p}\right)^2 \frac{1}{\tilde{\gamma}_0^2} \Gamma_L \qquad (3.48)$$

$$+ \frac{B(\vec{q})(\vec{q}\cdot\hat{Q})}{c_2 \int \frac{\mathrm{d}^3 q'}{(2\pi)^3} B(\vec{q}')(\vec{q}'\cdot\hat{Q})^2} \left(\frac{\omega v_E}{c_p}\right) \frac{\Gamma_L}{\tilde{\gamma}_0} \int \frac{\mathrm{d}^3 q}{(2\pi)^3} (\vec{q}\cdot\hat{Q}) \frac{1}{2} \frac{\Delta G_{\vec{q}}^2 (\vec{q}\cdot\hat{Q}) Q}{\tilde{\gamma}_0 \Delta G_0}.$$

On the other hand, the $\vec{q}'$ integrated two-particle Green's function is also known exactly within the ladder approximation and given by Eq. (3.40), so that we can continue from above

$$\int \frac{\mathrm{d}^3 q'}{(2\pi)^3} \Phi_{\vec{q}\vec{q}'} = \frac{\Delta G_{\vec{q}}}{\tilde{\gamma}_0^2 \Delta G_0} \Gamma_L + \frac{1}{2} \frac{\Delta G_{\vec{q}}^2 (\vec{q}\cdot\hat{Q}) Q}{\tilde{\gamma}_0^2 \Delta G_0} \Gamma_L. \qquad (3.49)$$

All what is left to do, is to compare coefficients of the right hand side of Eq. (3.48) with the ones occurring in Eq. (3.49) to find for the expansion coefficients

$$A(\vec{q}) = \Delta G_{\vec{q}} \qquad B(\vec{q}) = \Delta G_{\vec{q}}^2 \qquad (3.50)$$

and for the normalization constants also introduced in Eq. (3.47) we find

$$c_1 = \left(\frac{\omega}{c_p}\right)^2 \qquad c_2 = \left(\frac{\omega v_E}{c_p}\right). \qquad (3.51)$$

Employing those expressions for the expansion coefficients, one may eventually express the two-particle correlator $\Phi_{\vec{q}\vec{q}'}$ in the following way

$$\int \frac{\mathrm{d}^3 q'}{(2\pi)^3} \Phi_{\vec{q}\vec{q}'} = \frac{\Delta G_{\vec{q}}}{\left(\frac{\omega}{c_p}\right)^2 \int \frac{\mathrm{d}^3 q'}{(2\pi)^3} \Delta G_{\vec{q}'}} P_E^\omega(\vec{Q},\Omega) \qquad (3.52)$$

$$+ \frac{\Delta G_{\vec{q}}^2 (\vec{q}\cdot\hat{Q})}{\left(\frac{\omega v_E}{c_p}\right) \int \frac{\mathrm{d}^3 q'}{(2\pi)^3} \Delta G_{\vec{q}'}^2 (\vec{q}'\cdot\hat{Q})^2} J_E^\omega(\vec{Q},\Omega).$$

The above expression, Eq. (3.52), represents the complete expansion of the intensity correlator into its moments. This will be used in the next subsection to decouple the Bethe-Salpeter equation in momentum space and therefore to provide a solution of this equation for the photonic correlator.

3.5.3 General Solution of the Bethe-Salpeter Equation

The disorder averaged intensity correlation, the two-particle Green's function, obeys the Bethe-Salpeter equation, see Eq. (3.41)

$$\Phi_{\vec{q}\vec{q}'} = G_{q+}^{\omega+} G_{q-}^{*\,\omega-} \left[1 + \int \frac{\mathrm{d}^3 q''}{(2\pi)^3} \gamma_{qq''} \Phi_{\vec{q}''\vec{q}'}\right]. \qquad (3.53)$$

As already discussed, the Bethe-Salpeter equation may be rewritten into the kinetic or Boltzmann equation given in Eq. (3.42)

$$\left[\omega\Omega 2\mathrm{Re}\,\epsilon - Q\left(\vec{q}\cdot\hat{Q}\right) + \Delta\Sigma - \omega^2 \Delta\epsilon\right] \Phi_{\vec{q}}$$

$$= \Delta G_{\vec{q}} + \int \frac{\mathrm{d}^3 q'}{(2\pi)^3} \Delta G_{\vec{q}} \gamma_{\vec{q}\vec{q}'} \Phi_{\vec{q}'}. \qquad (3.54)$$

3.5. THEORY OF TRANSPORT AND LOCALIZATION

To find the solution of Eq. 3.54), in a first step one sums the equation over momenta $\vec{q}$, incorporates the generalized Ward identity as given in Eq. (3.43) and subsequently expands the obtained result for small internal momenta Q and internal frequencies Ω. Furthermore, it is essential to also employ the decoupling shown in Eq. (3.52). Eventually, after some straight forward algebraic manipulations, the generalized continuity equation for the energy density is found to be

$$\Omega P_{\rm E}^\omega - Q J_{\rm E}^\omega = \frac{4\pi i\,\omega\,N(\omega)}{g_\omega^{(1)}\left[1+\Delta(\omega)\right]c_p^2} \; + \; \frac{\vec{q}[g_\omega^{(0)}+\Lambda(\omega)]}{g_\omega^{(1)}\left[1+\Delta(\omega)\right]} P_{\rm E}^\omega \tag{3.55}$$

which represents energy conservation in the presence of optical absorption.

Within the standard solution procedure, above outlined, the next step is to obtain a linearly independent equation which also relates the energy density $P_{\rm E}^\omega$ and the current density $J_{\rm E}^\omega$. This is being done in a similar way as the one leading to Eq. (3.55): foregoing one first multiplies the kinetic equation, Eq.(3.54), by the projector $[\vec{q}\cdot\hat{Q}]$ and then follows the above outlined recipe to eventually obtain the wanted second relation, this is the so-called current relaxation equation

$$\left[\Omega\frac{{\rm Re}\epsilon_b}{c^2} + \frac{i}{c^2\tau^2} + iM(\Omega)\right] J_{\rm E}^\omega \; + \; \tilde{A}\,Q P_{\rm E}^\omega = 0 \;, \tag{3.56}$$

relating energy density $P_{\rm E}^\omega$ and energy density current $J_{\rm E}^\omega$ as required and furthermore introduces the so-called memory function $M(\Omega)$ according to

$$M(\Omega) = \frac{i\int\frac{{\rm d}^3q}{(2\pi)^3}\int\frac{{\rm d}^3q'}{(2\pi)^3}\;[\vec{q}\cdot\hat{Q}]\Delta G_{\vec{q}}^\omega \gamma_{\vec{q}\vec{q}'}^\omega (\Delta G_{\vec{q}'}^\omega)^2 [\vec{q}'\cdot\hat{Q}]}{\int\frac{{\rm d}^3q}{(2\pi)^3}\;[\vec{q}\cdot\hat{Q}]^2(\Delta G_{\vec{q}}^\omega)^2}. \tag{3.57}$$

where $\gamma_{\vec{p}\vec{p}'}^\omega \equiv \gamma_{\vec{p}\vec{p}'}^\omega(\vec{Q},\Omega)$ is the total irreducible two-particle vertex, which will be discussed in more detail in the following subsection. The memory kernel $M(\Omega)$ represents the system's response to the diffusing light intensity in the presence of absorption/gain.

So far, two independent equations, Eq. (3.55) and Eq. (3.56), have been obtained, both of them relating the current density $J_{\rm E}^\omega$ and density $P_{\rm E}^\omega$. Therefore one may now eliminate one of the two variables in this linear system of equations. One chooses to combine the two equations to find an expression for the energy density

$$P_{\rm E}^\omega(Q,\Omega) = \frac{4\pi i N(\omega)/(g_\omega^{(1)}\left[1+\Delta(\omega)\right]c_p^2)}{\Omega + iQ^2 D + i\xi_a^{-2}D}\;, \tag{3.58}$$

exhibiting the expected diffusion pole structure for non-conserving media, i.e. in the denominator of Eq. (3.58) there appears an additional term as compared to the case of conserving media. This is the term $\xi_a^{-2}D$, sometimes referred to as the mass term, accounting for loss (or gain) to the intensity not being due to diffusive relaxation. In Eq. (3.58) also the generalized, Ω-dependent diffusion coefficient $D(\Omega)$ has been introduced via the relation

$$D(\Omega)\left[1 - i\,\Omega\omega\tau^2{\rm Re}\epsilon_b\right] = D_0^{tot} - c^2\tau^2 D(\Omega) M(\omega). \tag{3.59}$$

Furthermore, Eq. (3.58) also introduces the absorption induced absorption length scale ξ_a of the diffusive modes,

$$\xi_a^{-2} \;=\; \frac{r_\epsilon A_\epsilon - 2\omega^2 {\rm Im}\epsilon_b}{2{\rm Re}\epsilon_b - A_\epsilon B_\epsilon/\omega}\frac{1}{\omega D(\Omega)}, \tag{3.60}$$

which is to be distinguished from the single-particle or light wave amplitude absorption or amplification length. The diffusion constant without the discussed memory effects (see later in Eq. (3.59)), $D_0^{tot} = D_0 + D_b + D_s$, consists of the bare diffusion constant

$$D_0 = \frac{2v_{\rm E} c_p}{\pi N(\omega)}\int\frac{{\rm d}^3q}{(2\pi)^3}\;[\vec{q}\cdot\hat{Q}]^2 ({\rm Im}\,G_{\vec{q}}^\omega)^2. \tag{3.61}$$

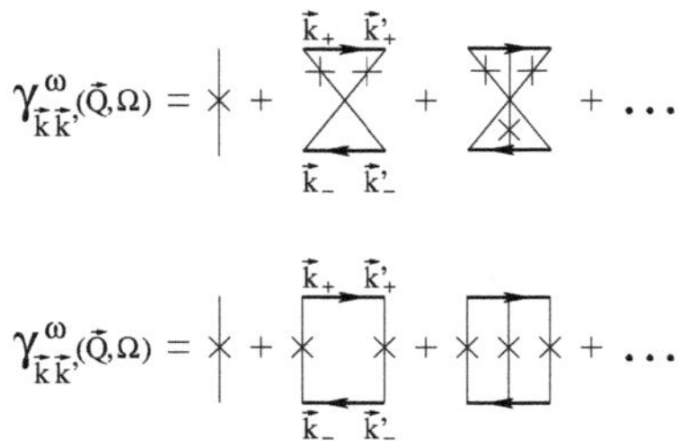

CHAPTER 3. LIGHT TRANSPORT IN INFINITE THREE DIMENSIONAL DISORDERED MEDIA WITH ABSORPTION OR AMPLIFICATION

$$\gamma^{\omega}_{\vec{k}\vec{k}'}(\vec{Q},\Omega) = \times + \underset{\vec{k}_-\ \vec{k}'_-}{\overset{\vec{k}_+\ \vec{k}'_+}{\bowtie}} + \underset{}{\bowtie} + \ldots$$

$$\gamma^{\omega}_{\vec{k}\vec{k}'}(\vec{Q},\Omega) = \times + \underset{\vec{k}_-\ \vec{k}'_-}{\overset{\vec{k}_+\ \vec{k}'_+}{\times\quad\times}} + \times\times\times\times + \ldots$$

Figure 3.3: *The upper panel shows a diagrammatic expansion of the irreducible two-particle vertex γ. The lower panel displays the disentangled Cooperon with changed momentum arguments as discussed in the text below.*

and renormalizations from absorption in the background medium (D_b) and in the scatterers (D_s),

$$D_b = (\omega\tau)^2 \, \mathrm{Im}\epsilon_b \, \tilde{D}_0/4 \, , \qquad D_s = r_\epsilon A_\epsilon \tau^2 \tilde{D}_0/8 \, , \tag{3.62}$$

where $\tilde{D}_0$ is the same as in Eq. (3.61), with $(\mathrm{Im}G^{\omega}_{\vec{q}})^2$ replaced by $\mathrm{Re}(G^{\omega}_{\vec{q}}{}^2)$ and consequently given by

$$\tilde{D}_0 = \frac{2v_E c_p}{\pi N(\omega)} \int \frac{\mathrm{d}^3 q}{(2\pi)^3} \, [\hat{q}\cdot\hat{Q}]^2 (\mathrm{Re}\, G^{\omega}_{\vec{q}})^2. \tag{3.63}$$

3.5.4 Vertex Function and Self-Consistency

From equations Eq. (3.57) and Eq. (3.59) it is clear that the energy density or two-particle function given in Eq. (3.58) still depends on the full two-particle vertex $\gamma^{\omega}_{\vec{q}'\vec{q}}$.
One may carefully analyze the vertex $\gamma^{\omega}_{\vec{q}'\vec{q}}$ for the self-consistent calculation of $M(\Omega)$ [39, 50], exploiting time reversal symmetry of propagation in the active/absorbing random medium. In the long-time limit ($\Omega \to 0$) the dominant contributions to $\gamma^{\omega}_{\vec{q}'\vec{q}}$ are the same maximally crossed diagrams (Cooperons) as for conserving media, which may also be disentangled. In Fig. (3.3) the disentangling of the Cooperon into the regular diffusion ladder is demonstrated. The internal momentum argument of the disentangled irreducible vertex function in the second line of Fig. (3.3) is replaced by the new momentum $\vec{Q} = \vec{k} + \vec{k}'$. By the described procedure $\gamma^{\omega}_{\vec{q}'\vec{q}}$ now acquires the absorption induced decay rate $\xi_a^{-2}D$. Finally the memory kernel $M(\Omega)$ reads

$$M(\Omega) = -\frac{(2v_E c_p)^2 \, u_\epsilon \left[2\pi\omega u_\epsilon N(\omega) + r_\epsilon A_\epsilon - 2\omega^2 \mathrm{Im}\epsilon_b\right]}{\pi\omega N(\omega) D_0 D(\Omega)} \tag{3.64}$$
$$\times \int \frac{\mathrm{d}^3 q}{(2\pi)^3} \int \frac{\mathrm{d}^3 q'}{(2\pi)^3} \, \frac{[\hat{q}\cdot\hat{Q}]|\mathrm{Im}G_q|(\mathrm{Im}G_{q'})^2\,[\hat{q}'\cdot\hat{Q}]}{\frac{-i\Omega}{D(\Omega)} + (\vec{q}+\vec{q}')^2 + \xi_a^{-2}}.$$

Eqs. (3.59)-(3.64) constitute the self-consistency equations for the diffusion coefficient $D(\Omega)$ and the decay length ξ_a in presence of absorption. Once the dynamic diffusion constant $D(\Omega)$ has been numerically evaluated, the full solution of the Bethe-Salpeter equation is established via Eqs. (3.64) and (3.58).
Therefore the initial problem of light intensity transport in infinite disordered media with absorption or gain including strong localization effects has to be regarded as solved.

3.6 Results and Discussion of the Transport Theory in Infinite 3-D Media with Absorption and Gain

In this chapter we have performed a semi-analytical theory for scalar waves propagating in three dimensional, random, dissipating, i.e. absorbing/emitting media. In Eqs. (3.11)-(3.16) we have

3.6. RESULTS AND DISCUSSION OF THE TRANSPORT THEORY IN INFINITE 3-D MEDIA WITH ABSORPTION AND GAIN

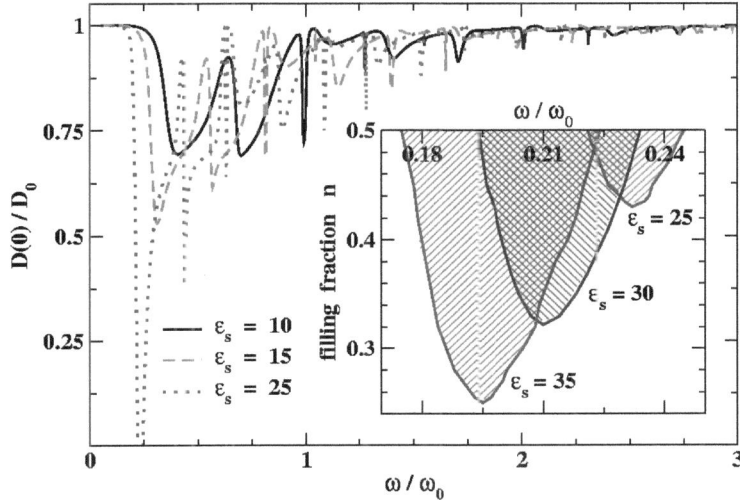

Figure 3.4: *Normalized diffusion constant $D(0)$ for $\epsilon_b = 1$ and scatterer volume fraction $\nu = 50\%$. The frequency unit in all figures is $\omega_0 = 2\pi c/r_0$ with r_0 the scatterer radius. The localization transition is reached for $\epsilon_s \gtrsim 24$. Inset: phase diagram in the (ν, ω) plane within the first Mie resonance for $\epsilon_b = 1$ and various ϵ_s. The shaded areas represent the localized phases.*

shown how stimulated emission (absorption) enters the transport within a semiclassical theory for infinite random media. As a consequence, finite emission or absorption rates induce additional renormalization terms to the dynamic diffusion constant as given in Eq. (5.92). Furthermore, we presented a detailed analysis of the solution of the Bethe-Salpeter equation, Eq. (3.41) for photonic correlations in the hydrodynamic, i.e. diffusive, limit. In this section we want to discuss in detail the influence of localization effects on intensity transport, included via a diagrammatic approach by generalizing the theory of Vollhardt and Wölfle. We put the focus on the calculated self-consistent equation for the dynamic diffusion constant $D(\Omega)$ in Eq. (3.59).

3.6.1 Discussion Transport Theory

Let us first discuss the kernel $M(\Omega)$. It describes the enhanced backscattering. For a conserving medium (Imϵ_b = Imϵ_s = 0), where $\xi_a^{-2} = 0$, it drives the Anderson Localization (AL) transition due to its negative, infrared divergent contribution [39, 35]. For illustration we show in Fig. 3.4 the AL phase diagram and $D(\Omega = 0)$ for $\epsilon_b = 1$ and various real values of ϵ_s, displaying strong suppression near the Mie resonances, albeit for rather high dielectric contrast $\Delta\epsilon$. Introducing now absorption or gain, the quadratic inverse correlation length ξ_a^{-2} becomes non-zero and can assume both positive and negative values. True AL due to repeated enhanced backscattering is no longer possible in this case, because, through self-consistency, a vanishing $D(0)$ would drive $\xi_a^{-2} \to \infty$ and $D(0)M(0) \to 0$. Inspection of $M(\Omega)$ shows analytically that to leading order in ξ_a^{-2} the real part of the integral in Eq. (3.64) is symmetrical with respect to $\xi_a^{-2} \leftrightarrow -\xi_a^{-2}$, i.e. AL is suppressed by absorption or gain in a symmetrical way. This is in agreement with the surprising absorption/gain duality of Ref. [43], which is valid for small gain and short times.

However, we do find important deviations from this result for systems where there is no symmetry between background and scattering medium (like in our three dimensional case): (1) As seen from

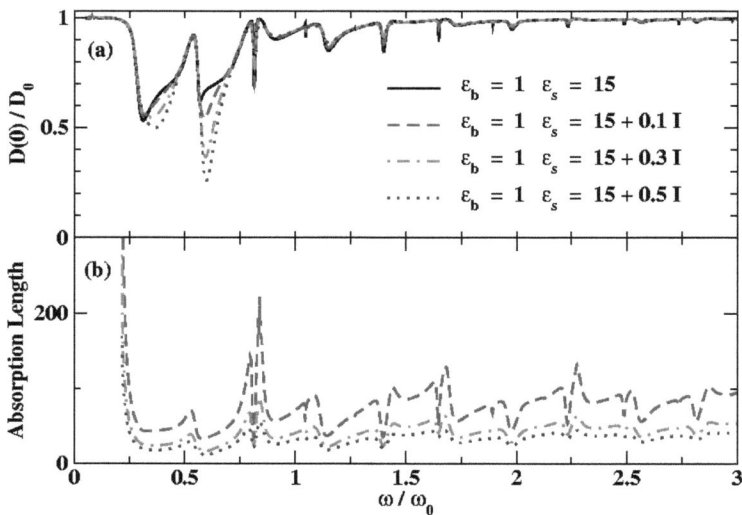

Figure 3.5: *(a) Diffusion constant $D(0)$ for $\nu = 50\%$. The scatterers exhibit different amounts of absorption as indicated. (b) Absorption length ξ_a in units of the scatterer radius for the systems in (a).*

Eq. (3.60), ξ_a^{-2} itself is not symmetric in the sign of $\text{Im}\epsilon_b$ or $\text{Im}\epsilon_s$ and can be positive even for purely emissive media, $\text{Im}\epsilon_b < 0$, $\text{Im}\epsilon_s < 0$. (2) The full diffusion coefficient has additional contributions D_b, D_s from loss/gain in the background and in the scatterers, Eqs. (5.92). While D_b is always positive for absorption and negative for gain as expected, D_s has a complicated dependence on the signs of $\text{Im}\epsilon_b$, $\text{Im}\epsilon_s$ and depends sensitively on whether absorption and/or gain occurs in the background medium or in the scatterers. This is because, in contrast to D_b, the impurity scattering contribution D_s results from an intricate interplay between elastic momentum relaxation and absorption/gain processes. We emphasize that the existence and the form of D_b, D_s, and ξ_a^{-2} are a direct consequence of the non-conserving terms in the WI Eq. (3.43), and, thus, are exact.

The complete scenario of localization effects is now as follows. Even though in the presence of absorption or gain there are no true Anderson localized modes because of the finite ξ_a^{-2}, the contributions D_b and D_s can strongly suppress the total diffusion coefficient $D(0)$. This is shown in Fig. 3.5 for a system of absorbing scatterers embedded in air. Even moderate absorption drastically decreases the diffusion constant $D(0)$ close to the low-order Mie resonances. At the same frequencies the correlation length ξ_a is suppressed even more dramatically. For the case of a purely absorbing system, ξ_a may be identified with the effective absorption length for diffusive modes. For the case of emission, e.g. in the background, two scenarios are possible. (i) $\xi_a^{-2} > 0$; this may occur due to the subtle interplay with momentum relaxation processes and absorption in the scatterers. Both $D(0)$ and the intensity correlation length ξ_a are real and finite, but suppressed near the Mie resonances, as shown in Fig. 3.6. It is also seen that, e.g., absorption in the scatterers can be partially compensated by emission in the background. (ii) Re $\xi_a^{-2} < 0$; this is realized for sufficiently strong gain. It implies a pole on the real axis in the integration range of Eq. (3.64) and, hence, complex $D(0)$ and ξ_a. Fourier transforming Eq. (3.58), this means an exponential intensity growth for long times with rate $1/\tau_a = \text{Re}[(\xi_a^{-2} + Q^2)D(0)]$, in qualitative agreement with Ref. [26], thus reconciling these long-time results with the weak gain or short-time results of Ref. [43]. This growth is modulated by temporal and spatial oscillations with characteristic frequency $\Omega_D = -Q^2 \text{Im}D(0)$ and wave number

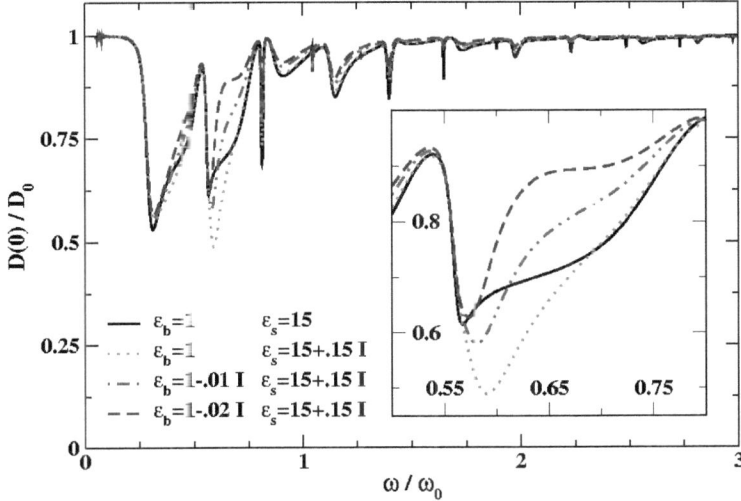

Figure 3.6: *Diffusion constant $D(0)$ for absorbing scatterers ($\nu = 50\%$) in a host medium with optical gain, as indicated. Parameters are chosen such that $\xi_1^2 > 0$.*

$k_D = \text{Im}(1/\xi_a)$, respectively. We interpret this oscillatory behavior as a memory effect, originating from the competition between the enhanced backscattering of waves and their amplified propagation in the surrounding medium. Self-induced oscillations have been found before in other driven systems with competing dynamics.

3.6.2 Causality and Length Scales

The second issue we want to point out, is concerned with the pole structure of the energy density correlator in Eq. (3.58). As can be seen from Eq. (3.60), the square of the correlation length may become negative for amplifying media. This raises the question of causality of P_E. Causality means here that the pole (see Fig. ??) of P_E in the complex frequency plane must reside in the lower half of that plane. Since the value of ξ_a is determined once the system has been specified, causality sets a constraint of the allowed Q- modes in that system.
Generally this is an amazing remark, as e.g. in a simpler system of a single microsphere with gain, it has been shown experimentally [27], that the scattering coefficients calculated within linear response lose their causality exactly at the point where the sphere crosses its lasing threshold. So in turning this argument around, we can also conclude, that linear response theory predicts the laser threshold of the sphere correctly. Technically this is seen in the behavior of the single particle Green's function. The single particle Green's function looses it's causality when the self energy Σ (see Fig. ??) looses its causality. Although, again, linear response theory is of course not capable of describing the lasing regime itself.
In particular there exists a minimum momentum Q_{min}, and only diffusive modes with $Q \geq Q_{min}$ exist in the system. The corresponding length scale is then given by $R_{max} = 2\pi\sqrt{\xi^2 \text{Re}(D)/D}$ and defines the maximum length over which diffusive modes can be correlated. This length may be experimentally measured as the spot size of a lasing mode [53].
Although this transport theory with linear gain is not suited to describe lasing, it still provides deeper insight in light propagation in gain media and even predicts a maximum correlation volume,

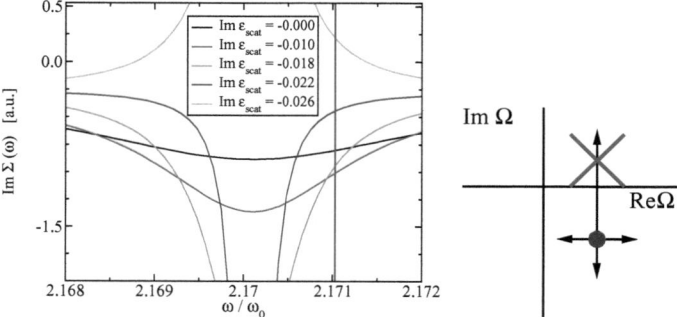

Figure 3.7: *Left panel: The behavior of the imaginary part of the self energy* Σ *of the single particle Green's function is displayed as a function of the light frequency for several increasing gain values. In the frequency window around a Mie-resonance we see that for increasing gain the resonance of* ImΣ *narrows and becomes deeper. Finally the width vanishes and then* ImΣ *changes the sign. This means the causality threshold is reached.*
Right panel: Causality requires that the diffusion pole remains in the lower half of the complex plane. This demands a new maximum length scale R_{max}.

i.e. a maximum spot size, for laser active media.
A numerical evaluation of the here derived correlation length R_{max} is presented in Fig. 3.8 as a function of Imϵ_s, which serves here as a measure of the external pumping strength. A comparison of the calculated and measured spot size shows a good qualitative agreement.
With these promising results one is encouraged to take the next step, and consider a random laser in the diffusive regime, characterized by a diffusion constant D_0. For the moment Cooperon contributions are neglected.

3.7 Conclusion

In conclusion, we have presented in this chapter a semi-analytical theory for the interplay of strong localization effects and absorption or gain. True AL is not possible in the presence of either loss or gain. However, strong renormalizations of the diffusion constant D arise from the violation of the conservation laws. These renormalizations depend sensitively on whether absorption/gain occurs in the scatterers or the background. Intimately connected with the suppression of D is the appearance and reduction of a finite intensity correlation length ξ_a, even though there are no truly localized modes. ξ_a includes effects of both, impurity scattering and loss/gain in the medium, and, thus, can be shorter than the scattering mean free path l. For example, in a pure medium with loss/gain ξ_a would characterize the absorption/gain length and would be finite, while l would obviously be infinite. It should be emphasized that the present theory incorporates both, the physics of multiple impurity scattering valid at length scales larger than l (Diffuson and Cooperon contributions) and the physics of coherent amplification present at all length scales. Therefore, it is expected to give an accurate estimate for ξ_a. We conjecture that in random lasers this finite length scale ξ_a might define the coherence volume necessary for resonant feedback, that is observed experimentally [45] and that appears to be smaller than l in those experiments. In order to substantiate this conjecture, the present localization theory will be coupled self-consistently to the laser rate equations in the following chapter, which yields results for the position dependent dielectric function above the lasing threshold (Im$\epsilon_{b,s} \neq 0$) and, hence, for the lasing mode volume. Oscillatory behavior in space and time is also predicted for sufficiently strong gain, although it is presumably difficult to observe,

3.7. CONCLUSION

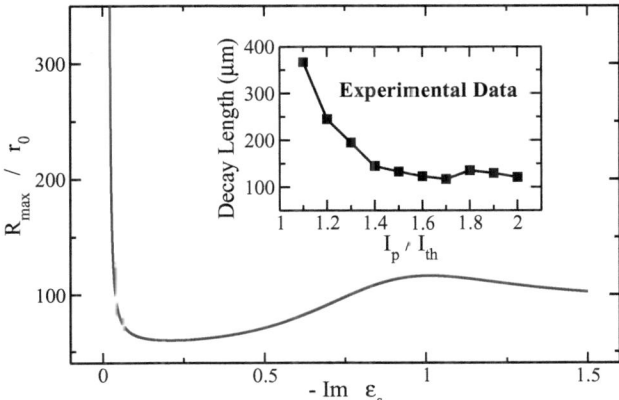

Figure 3.8: *The correlation length R_{max} is shown as predicted from the diffusion pole of the energy density correlator as a function of increasing imaginary part of scatterers, which is proportional to the external pumping. The following parameters have been used: $\epsilon_b = 1$, $\text{Re}\epsilon_s = 10$, the filling fraction of the scatterers $\nu = 30\%$, the external light frequency $\omega/\omega_0 = 2.5$, where $\omega_0 = 2\pi c/r_0$ and c is the vacuum speed of light. R is normalized to the scatter radius r_0. The experimental data in the inset are taken from ref. [53] and refer to the spot size of the modes. The calculated and measured spot size show a good qualitative agreement.*

as it occurs only during the exponential intensity growth between the laser threshold and saturation.

Chapter 4

Phenomenological Model of a Random Lasing Layer

In the preceeding chapter we developed the transport theory for random, laser-active systems, including light diffusion and weak localization. In the current chapter we show the coupling of a purely diffusive model to the semiclassical laser rate equations, and further that the surface of the laser-active medium is crucial in order to stabilize a stationary lasing state. We solve the laser transport theory for bulk material with appropriate surface boundary conditions to obtain the spatial distributions of the light intensity and of the occupation inversion. The dependence of the intensity correlation length on the pump rate agrees with experimental findings. This chapter marks an intermediate step towards a microscopic random lasing theory.

4.1 Introduction

As we have already discussed in this thesis before, a random laser is a system formed by stochastically distributed scatterers embedded in a host medium, if either scatterer or host medium or both provide optical gain. Multiple scattering of light is the mechanism providing coherent laser feedback, which has been demonstrated experimentally to be present in such systems beyond doubt [51, 52, 53, 54, 55, 56, 57]. The origin of those spatially confined photonic quasi-modes, required by the coherent feedback, has remained controversial. To gain a better understanding of the interplay of optical gain and effects due to strong or Anderson localization (AL) [64], the investigation of light transport in (linear-) gain media is an appropriate point to begin with in this chapter.

The underlying diffusion processes including AL, which itself has been understood [59, 60] as an effect of repeated self-interference (so-called Cooperon contributions) of diffusive modes, is a hydrodynamic phenomenon. Therefore it relies on conservation laws, which are, in simple terms, broken in a gain medium. A theory including amplification in terms of linear gain, i.e. as an imaginary part of the dielectric constant, is described in the chapter before and provides the needed ground work for the actual random lasing systems.

Here we discuss the question of how a random laser system in the diffusive regime can be described. To this end we consider the lasing material itself by the semi-classical laser rate equations and combine it with a diffusing light intensity as described by a diffusion equation. Intensity loss introduced by surfaces of finite disordered gain media is found to be crucial in establishing a stationary behavior of lasing modes.

4.2 Theory of a Diffusive Random Laser

We propose a theory of a diffusive random laser in the stationary regime. The lasing properties of the medium are described by the semi-classical laser rate equations, which read for a four level laser

$$\frac{\partial N_3}{\partial t} = \frac{N_0}{\tau_P} - \frac{N_3}{\tau_{32}} \tag{4.1}$$

$$\frac{\partial N_2}{\partial t} = \frac{N_3}{\tau_{32}} - \left(\frac{1}{\tau_{21}} + \frac{1}{\tau_{nr}}\right) N_2 - \frac{(N_2 - N_1)}{\tau_{21}} n_{ph} \tag{4.2}$$

$$\frac{\partial N_1}{\partial t} = \left(\frac{1}{\tau_{21}} + \frac{1}{\tau_{nr}}\right) N_2 + \frac{(N_2 - N_1)}{\tau_{21}} n_{ph} - \frac{N_1}{\tau_{10}} \tag{4.3}$$

$$\frac{\partial N_0}{\partial t} = \frac{N_1}{\tau_{10}} - \frac{N_0}{\tau_P} \tag{4.4}$$

$$N_{tot} = N_0 + N_1 + N_2 + N_3, \tag{4.5}$$

where N_i are the population number of the corresponding electron level ($i \in \{1\ldots 4\}$), N_{tot} is the total number of electrons, $\gamma_{ij} \equiv 1/\tau_{ij}$ are the transition rates from level i to j and $\gamma_P \equiv 1/\tau_P$ is the transition rate due to homogeneous constant external pumping. Furthermore $n_{ph} \equiv N_{ph}/N_{tot}$ is the relative photon number or photon density. Combining the above five equations in the stationary limit ($\partial_t N_i = 0$), assuming γ_{32} and γ_{10} to be very large, the population inversion is found to be

$$n_2 = \frac{\gamma_P}{\gamma_P + \gamma_{nr} + \gamma_{21}(n_{ph} + 1)}, \tag{4.6}$$

where n_2 is defined as $n_2 \equiv N_2/N_{tot}$. As mentioned above, we seek to describe a diffusive random laser, so that photon density in the disordered medium is required to obey the diffusion equation

$$\partial_t n_{ph} = D_0 \nabla^2 n_{ph} + \gamma_{21}(n_{ph} + 1)n_2, \tag{4.7}$$

where the last term of the r.h.s. describes the intensity increase due to stimulated and spontaneous emission, as found in the semi-classical laser rate equations. In the stationary limit the diffusing intensity therefore obeys

$$\nabla^2 n_{ph} = -\frac{\gamma_{21}}{D_0}(n_{ph} + 1)n_2. \tag{4.8}$$

Since in this section we are not concerned with correlation functions but rather with intensity, i.e. the relative photon number n_{ph}, an infinite medium cannot be considered, as it would inevitably lead to an infinite photon number. Therefore we consider a system which is bounded in the $z-$ direction, $-\frac{d}{2} \leq z \leq \frac{d}{2}$, and infinite otherwise. This symmetry introduces a loss of intensity since photons may leave the disordered gain medium through the two surfaces. As it will be shown below this introduced loss is crucially needed to stabilize the lasing mode allowing for a stationary solution. In this film geometry a Fourier transform with respect to the unbounded variables x, y may be defined in the usual way. Applying this Fourier transformation to e.g. $\nabla^2 n_{ph}(x, y, z)$ leads to $(-Q_{||}^2 + \partial_z^2) n_{ph}(Q_{||}, z)$, where $Q_{||}^2 = Q_x^2 + Q_y^2$, is the momentum parallel to the film. Using this transformation, the stationary limit of the photon number may be re-written as

$$n_{ph}(x, y, z) = \int \frac{dQ_{||}^2}{(2\pi)^2} e^{-i\vec{Q}_{||} \cdot \vec{r}_{||}} P_E(\vec{Q}_{||}, z) n_2(\vec{Q}_{||}, z), \tag{4.9}$$

with the energy density kernel $P_E(\vec{Q}_{||}, z)$ given as

$$P_E(\vec{Q}_{||}, z) = \frac{\gamma_{21}}{-i\Omega + D_0 Q_{||}^2 + D_0 \xi^{-2}}, \tag{4.10}$$

where the correlation length ξ is defined as

$$\xi = \sqrt{\frac{D_0}{\gamma_{21}} \frac{n_{ph}}{n_2}}. \tag{4.11}$$

4.2. THEORY OF A DIFFUSIVE RANDOM LASER

In order to obtain Eq. (4.10) we used the translational invariance of the system in the $(x,y)-$plane and explicit diffusive behavior in the $z-$ direction by means of the one dimensional stationary diffusion equation in $z-$ direction

$$D_0 \partial_z^2 n_{ph} = -\gamma_{21}(n_{ph} + 1)n_2. \tag{4.12}$$

As seen in Eq. (4.10), the pole structure in this purely diffusive model behaves uncritical with respect to causality, as expected. The correlation length ξ, and thus also the mass term $D_0 \xi^{-2} = \gamma_{21} \frac{n_2(\vec{Q}_{||},z)}{n_{ph}(\vec{Q}_{||},z)}$, always remains positive, indicating an effective loss out of a given $Q_{||}$ mode. Additionally, the mass term becomes less and less significant as the laser intensity in the sample builds up. This is so because the relative population inversion clearly obeys $n_2 \leq 1$ whereas the relative photon number is not restricted.

To obtain explicit results it is therefore sufficient to combine Eqs. (4.12) and (5.99) to obtain

$$\partial_z^2 n_{ph}(z) = -\frac{\gamma_{21}}{D_0} \frac{(\gamma_P/\gamma_{21})}{1 + \frac{(\gamma_P/\gamma_{21})}{n_{ph}(z)+1}}, \tag{4.13}$$

describing the intensity (relative photon number) of a random laser film as function of the $z-$direction, perpendicular to the film's surface. Due to the translational invariance of the system in the $(x,y)-$ plane, the solution of Eq. (4.13) applies to any position within the plane of the lasing material.

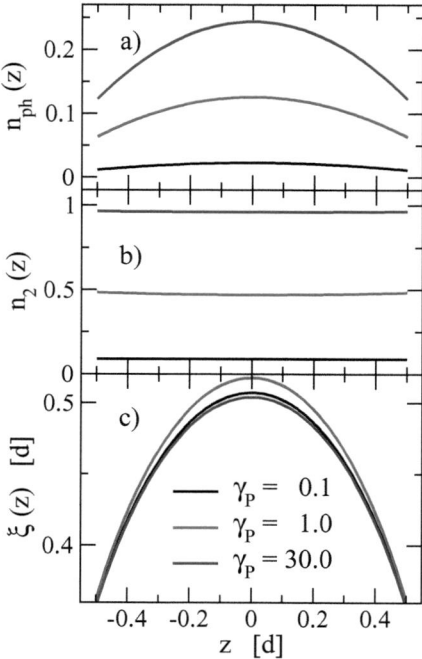

Figure 4.1: *Shown is the behavior of relevant quantities as a function z, the direction perpendicular to the surface, through the film of the random lasing material for different values of the pumping rate γ_P measured in units of γ_{21}. The diffusion constant is $D_0 = 1d^2\gamma_{21}$. Panel a) displays the photon number. The photon number increases monotonically with increasing pumping, and has its maximum in the center of the film. Panel b) displays the population inversion of the lasing material, which is roughly inverse to the photon number. Panel c) however shows the correlation length of the lasing system, and clearly behaves non-monotonically with increasing pumping.*

4.3 Results and Discussion

Numerical evaluations of Eqs. (4.13), (4.11) and (5.99) are shown in Figs. 4.1 and 4.2. In Fig. 4.1 the photon number $n_{ph}(z)$, the population inversion $n_2(z)$ and also the correlation length of the lasing intensity $\xi(z)$ is shown as a function of z for different values of the external pumping, characterized by the pumping rate γ_P, measured in units of γ_{21}. The value of the diffusion constant was chosen to be $D_0 = 1d^2\gamma_{21}$, where d is the width of the film. In panel a) of Fig. 4.1 the photon number displays a monotonically increasing behavior with increasing pumping. The maximum of the intensity resides in the center of film ($z = 0$), since this is the position farthest from the boundaries, and therefore with lowest loss of intensity. The population inversion, see Eq. (5.99), behaves roughly inverse to $n_{ph}(z)$, and monotonically with increasing pumping. In contrast to this, the correlation length $\xi(z)$ as given by Eq. (4.11) displays a non-monotonic behavior with increasing pumping. For pumping rates $\gamma_P < \gamma_{21}$ the correlation length increases but for pumping rates $\gamma_P > \gamma_{21}$, ξ is decreasing. The equality between γ_P and γ_{21} marks the situation where electrons are as fast excited into the upper laser level as they relaxate to lower levels. Therefore this characterizes the lasing threshold. Since experimental data are so far available for systems above threshold only [53], only the decreasing behavior of the correlation length or spot size has been reported.

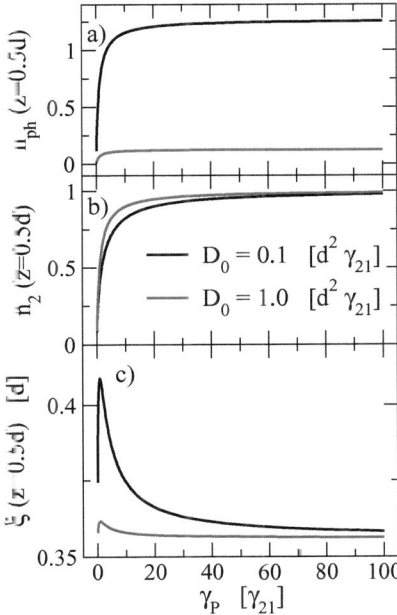

Figure 4.2: *Shown is the behavior of relevant quantities as a function of the pumping rate γ_P (measured in units of γ_{21}) at the surface of the film ($z = 0.5d$) for two different values of the diffusion constant D_0. Panel a) displays the photon number. The photon number reaches saturation at high pumping, the numerical value depends on the diffusion constant. Panel b) displays the population inversion of the lasing material, saturating towards full inversion ($n_2 = 1$). Panel c) however shows the correlation length of the lasing system, and clearly behaves non-monotonically with increasing pumping. Below the lasing threshold ($\gamma_P/\gamma_2 = 1$) the correlation length, i.e. the spot size, increases, and afterwards decreases again.*

In Fig. 4.2 the photon number $n_{ph}(z)$, the population inversion $n_2(z)$ and also the correlation length of the lasing intensity $\xi(z)$ is shown as a function of external pumping, characterized again by the pumping rate γ_P, measured in units of γ_{21}. The different quantities are evaluated at the surface of the lasing film, i.e. at $z = 0.5d$. Shown are results for two different diffusion constants $D_0 = 0.1d^2\gamma_{21}$ and $D_0 = 1d^2\gamma_{21}$, both of which are experimentally accessible in these systems. In panel a) of Fig. 4.2 the displayed photon number increases with increasing pumping until saturation behavior sets in. The exact value of the saturation intensity depends on the diffusion constant. If the diffusion constant is large, then the time a photon spends inside the material is small. If the diffusion is less favored, i.e. for smaller D_0, then the intensity remains longer inside the lasing material and therefore contributes more to the population inversion. Consequentially the built up intensity within the material is larger in the latter case. In panel b) the population inversion also increases with increasing pumping until it saturates towards full inversion $n_2 = 1$, i.e. all available electrons have been excited into the upper laser level. The correlation length ξ in panel c) however, displays a non-monotonic behavior, increasing below the threshold and decreasing above. The decreasing behavior above threshold has been reported in experimental data [53].

4.4 Summary

We presented in this chapter a coupling of the semi-classical laser rate equations to a disordered system of finite size, which is described by an assumed diffusion equation and losses through the surfaces.

Compared to chapter 3 we took the complementary point of view and treated the lasing properties correctly in terms of semi-classical laser rate equations but neglected Cooperon contributions and used the bare diffusion constant to describe a diffusive random laser. Of particular interest to us is at this point the stationary behavior where the constant external pumping and the emitted light intensity establish a balancing. For this case we showed that the corresponding energy density correlation function always behaves causal as it must. This is due to intensity loss through the samples surface, which therefore stabilize the lasing mode. The second major result in this section is the calculation of the correlation function, describing the spot size of the lasing modes. We find that above laser threshold the spot size decreases with increasing pumping in quantitative agreement with experimental data. Furthermore we find an increasing behavior with increasing pump intensity below the threshold.

Chapter 5

Selfconsistent Microscopic Theory of Random Lasing

In chapter three we developed a transport theory of light in bulk material, which is infinite in all three spatial dimensions and therefore also translational invariant. In experimental physics [63], however, the relevant geometries are always of finite size, e.g. the purely two dimensional systems of finite extension as used in experiments by H. Kalt [63], or, also of particular interest to us, the slab or film geometry as shown in Fig. 5.1, which is widely used in experiments by H. Cao [2]. In the latter system, the extension within the (x, y) - plane can be regarded as infinitely large compared with intrinsic length scales. Hence the system obtains translational invariance in this plane, the finite extension in z-direction, however, breaks this translational symmetry regarding the intensity transport in this third dimension.

The proposed phenomenological lasing theory with loss terms at the boundaries in the z-direction is already well designed to describe such a finite system, but it fails to account for transport details. Thus there is a need to adjust the transport theory for such systems, especially when the considered layers are thin, because then the boundary effects contribute the most.

In this chapter, we introduce a transport theory, which includes translational invariance in the (x, y) - plane and also accounts for the finite width in z-direction, c.f Fig. 5.1. We find a subtle interplay between both of these conditions when we look at the diffusive characteristics of the intensity propagator. As one consequence, the causality of any observable is intrinsically guaranteed. This will be shown to result from the light intensity loss through the sample's surface. If we look at the transport theory introduced in chapter three it can easily be seen, that introducing a finite length in one direction of the sample will cause significant modifications because this theory has been developed in the momentum and frequency domain. Due to the finite length applying a Fourier-transformation requires special attention.

What we have to derive first is a fully consistent description for the single-particle behavior, while assuming that the wave traveling through the sample experiences bulk properties of the material. Afterwards we will look at the two-particle quantities, obeying the Bethe-Salpeter-Equation. Here we have to deal with the problem, that the intensity sees the surface effects in the third direction. This results in a different treatment of the spacial dimensions in-plane and the finite z-direction. Nevertheless light intensity experience three dimensional transport.

5.1 Single - Particle Green's Function

In this chapter we introduce the intensity transport in a system which is translational invariant in-plane and of finite length in the third direction. In contrast to the theory for infinite systems in space we have to use some modifications for the slab geometry. Contrary to chapter three, where we performed a complete Fourier transformation of the quantities, we have to deal here with a different situation which forces us to reassess the Fourier transformation in the third direction. The finite extension in this direction prohibits that we may perform a Fourier transformation in this direction

 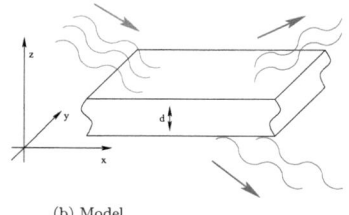

(a) Slab geometry (b) Model

Figure 5.1: *(a) The geometry of the considered system is depicted [1] . Within the (x,y) - plane the system can be regarded as infinitely large as compared to any intrinsic length scale, therefore the system displays translational invariance within this plane. The width of this film or slab geometry, however, is of finite width d and requires special consideration, as it introduces intensity losses at the surface. (b) Slab geometry model including translational invariance in-plane and finite extension in the third direction. Loss at the surface of the system balances the stationary limit.*

straight forwardly. Therefore we introduce a transformation of the z-coordinate, which is finite, to the center of mass and a relative coordinate. With respect to the relative coordinate the system behaves approximately translational invariant, we may therefore perform a Fourier transformation of this coordinate to momentum space. The transformation to center of mass and relative coordinates leads to the approbate expression for the Boltzmann or kinetic equation as will be shown below.

In order to find an explicit expression and also in order to apply a (partial) Fourier transformation we need to have a closer look at the single-particle-Green's function first.

As the starting point we chose the defining equation for the retarded disorder averaged single-particle Green's function

$$\left(\epsilon_b \omega_+^2 + \nabla_{r_1}^2 - \Sigma^R\right) G(r_1, r_1'; \omega_+) = \delta(r_1 - r_1'). \tag{5.1}$$

We now introduce center of mass and relative coordinates as

$$R = \frac{r_1 + r_1'}{2} \qquad\qquad r = r_1 - r_1' \tag{5.2}$$

and replace r_1 and r_1' and the gradient according to Eq. (E.3). After this, Eq. (5.1), is now rewritten as

$$\left(\epsilon_b \omega_+^2 + \frac{1}{4}\nabla_R^2 + \nabla_{\Delta r}^2 + \nabla_R \nabla_{\Delta r} - \Sigma^R\right) G(R, \frac{\Delta r}{2}; \omega_+) = \delta(\Delta r'). \tag{5.3}$$

With respect to relative coordinates, the system is approximately translational invariant, and a Fourier transformation can be applied. Applying such transformation to the above equation yields

$$\left(\epsilon_b \omega_+^2 + \frac{1}{4}\nabla_R^2 - p^2 + i\vec{p} \cdot \nabla_R - \Sigma^R\right) G_p(R; \omega_+) = \mathbb{1}. \tag{5.4}$$

In contrast to the relative coordinate the center of mass coordinate is translationally invariant in-plane only. Therefore the system can be Fourier-transformed with respect to the x- and y-components of the center of mass coordinate

$$\left(\epsilon_b \omega_+^2 - \frac{1}{4}Q_\parallel^2 + \frac{1}{4}\partial_Z^2 - p^2 - p_x Q_x - p_y Q_y + i p_z \partial_Z - \Sigma^R\right) G_p(\vec{Q}_\parallel; Z; \omega_+) = \mathbb{1} \tag{5.5}$$

5.1. SINGLE - PARTICLE GREEN'S FUNCTION

where we define $\vec{Q}_{||}$ to be a three-dimensional vector of the form

$$\vec{Q}_{||} = \begin{pmatrix} Q_x \\ Q_y \\ 0 \end{pmatrix}. \tag{5.6}$$

Then we rewrite the above differential equation to

$$\left(\left[\epsilon_b \omega_+^2 - \left(\vec{p} + \frac{\vec{Q}_{||}}{2}\right)^2 - \Sigma^R\right] + \frac{1}{4}\partial_Z^2 + ip_z\partial_Z\right) G_p(\vec{Q}_{||}; Z; \omega_+) = 1. \tag{5.7}$$

This differential equation, Eq. (5.7), has to be solved with Dirichlet boundary conditions. The general solution can readily be obtained as

$$G_p(\vec{Q}_{||}; Z; \omega_+) = \frac{1}{\epsilon_b \omega_+^2 - \left(\vec{p} + \frac{\vec{Q}_{||}}{2}\right)^2 - \Sigma^R} + C_1 e^{-2i\left(p_z + \sqrt{p_z^2 + a^2}\right)Z} + C_2 e^{-2i\left(p_z - \sqrt{p_z^2 + a^2}\right)Z} \tag{5.8}$$

where the abbreviation a is defined as

$$a := \left[\epsilon_b \omega_+^2 - \left(\vec{p} + \frac{\vec{Q}_{||}}{2}\right)^2 - \Sigma^R\right] \tag{5.9}$$

and the coefficients C_1 and C_2 have to be evaluated according to the boundary conditions. As boundary conditions we require that the Green's function takes some value R at both surfaces, i.e. at $Z = \pm\frac{d}{2}$, therefore

$$G_p(\vec{Q}_{||}; Z = \pm\frac{d}{2}; \omega_+) = R, \tag{5.10}$$

i.e. we impose symmetric boundary conditions and the value of R is to be determined below. R is not to be mixed up with the center of mass coordinate of the chapter before. Having specified these conditions, we can solve the system of equations (as shown in detail in Appendix B) for the two coefficients C_1 and C_2 and eventually obtain the result

$$C_1 = \left(R - \frac{1}{a}\right) \frac{\sin\left(p_z d + \sqrt{p_z^2 + a^2} \cdot d\right)}{\sin\left(\sqrt{p_z^2 + a^2} \cdot 2d\right)} \tag{5.11}$$

$$C_2 = \left(\frac{1}{a} - R\right) \frac{\sin\left(p_z d - \sqrt{p_z^2 + a^2} \cdot d\right)}{\sin\left(\sqrt{p_z^2 + a^2} \cdot 2d\right)}. \tag{5.12}$$

In order to determine the value of the Green's function at the two boundaries we now require, that the single-particle Green's function in the very vicinity of the surface has approximately bulk characteristics, i.e. we set $R = 1/a$. This is justified by the observation, that the Green's function describes the characteristics of a single wave, having characteristic length scale wavelength λ, which is much smaller than the system size d and also much smaller than the center of mass coordinate, or length scale Z. Therefore the Green's function varies only very slowly as a function Z, and this variation may as well be neglected since it will not enter the localization theory in any critical way. Finally, the solution for the single-particle Green's function is found to be

CHAPTER 5. SELFCONSISTENT MICROSCOPIC THEORY OF RANDOM LASING

Figure 5.2: *Diagrammatic form of the representation independet Bethe-Salpeter-equation of the intensity-correlation function Φ as shown in Eq. (5.14).*

$$G_p(\vec{Q}_{\|}; Z; \omega_+) = \frac{1}{\epsilon_b \omega_+^2 - \left(\vec{p} + \frac{\vec{Q}_{\|}}{2}\right)^2 - \Sigma^R} \qquad (5.13)$$

and is not a function of the center of mass coordinate Z. This is the expression of the single-particle Green's function which will be used further.

5.2 Light - Intensity Correlation Function

As we have seen, the position dependencies of our system has to be discussed in detail concerning the Fourier-transformation. For the reason that not all coordinates can be Fourier-transformed straight forward, we need the position-space representation of the Bethe-Salpeter equation (BS) to discuss the Fourier-transformation in detail. Therefore we start with the representation-independent operator notation of the BS, which reads

$$\hat{\Phi} = \left(\hat{\mathbf{G}}^R \otimes \hat{\mathbf{G}}^A\right) + \left(\hat{\mathbf{G}}^R \otimes \hat{\mathbf{G}}^A\right)\hat{\gamma}\hat{\Phi} = \left(\hat{\mathbf{G}}^R \otimes \hat{\mathbf{G}}^A\right)\left[\mathbb{1} \otimes \mathbb{1} + \hat{\gamma}\hat{\Phi}\right]. \qquad (5.14)$$

To find the real-space representation, we first construct a state in real space as

$$|r_1, r_2\rangle \equiv |r_1\rangle \otimes |r_2\rangle \qquad (5.15)$$

with meaning, that $|r_1\rangle$ acts only on retarded subspace and $|r_2\rangle$ acts only on advanced subspace. The corresponding bra is therefore

$$\langle r_1', r_2'| \equiv \langle r_1'| \otimes \langle r_2'| \qquad (5.16)$$

where again $\langle r_1'|$ acts only on retarded subspace and $\langle r_2'|$ acts only on advanced subspace.
So for instance the product for disorder-averaged Green's functions reads

$$\langle r_1, r_2|\hat{\mathbf{G}}^R \otimes \hat{\mathbf{G}}^A|r_1', r_2'\rangle = \langle r_1|\hat{\mathbf{G}}^R|r_1'\rangle \otimes \langle r_2|\hat{\mathbf{G}}^A|r_2'\rangle \qquad (5.17)$$
$$= G^R(r_1, r_1') \otimes G^A(r_2, r_2') \qquad (5.18)$$
$$= G^R(r_1, r_1')G^A(r_2, r_2') \qquad (5.19)$$

where $G^R(r_1, r_1')$ is the regular real-space representation of G, i.e. this is the solution of the defining differential equation.
In the following we will use the completeness relation of the states

$$\sum_{r_3, r_4} |r_3, r_4\rangle\langle r_3, r_4| = \sum_{r_3, r_4}(|r_3\rangle \otimes |r_4\rangle)(\langle r_3| \otimes \langle r_4|) \qquad (5.20)$$
$$= |r_3\rangle\langle r_3| \otimes |r_4\rangle\langle r_4| \qquad (5.21)$$
$$= \mathbb{1} \otimes \mathbb{1} = \mathbb{1}. \qquad (5.22)$$

5.2. LIGHT - INTENSITY CORRELATION FUNCTION

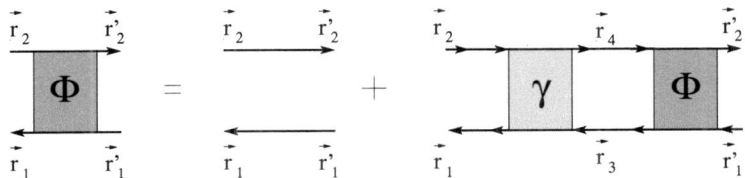

Figure 5.3: *Position-space representation of the Bethe-Salpeter-equation as shown in Eq. (5.25).*

Multiplying the BS equation with $\langle r_1, r_2|$ from the left and with $|r'_1, r'_2\rangle$ from the right we arrive at

$$\langle r_1, r_2| \hat{\Phi} |r'_1, r'_2\rangle = \langle r_1, r_2| \left(\hat{\mathbf{G}}^R \otimes \hat{\mathbf{G}}^A \right) |r'_1, r'_2\rangle \qquad (5.23)$$
$$+ \langle r_1, r_2| \left(\hat{\mathbf{G}}^R \otimes \hat{\mathbf{G}}^A \right) \hat{\gamma} \hat{\Phi} |r'_1, r'_2\rangle$$

and we insert two $\mathbb{1}$, i.e. $\mathbb{1} = \sum_{r_3, r_4} |r_3, r_4\rangle\langle r_3, r_4|$ and $\mathbb{1} = \sum_{r_5, r_6} |r_5, r_6\rangle\langle r_5, r_6|$ to obtain

$$\langle r_1, r_2|\hat{\Phi}|r'_1, r'_2\rangle = \langle r_1, r_2| \left(\hat{\mathbf{G}}^R \otimes \hat{\mathbf{G}}^A \right) |r'_1, r'_2\rangle \qquad (5.24)$$
$$+ \sum_{r_3, r_4} \sum_{r_5, r_6} \langle r_1, r_2| \left(\hat{\mathbf{G}}^R \otimes \hat{\mathbf{G}}^A \right) |r_5, r_6\rangle\langle r_5, r_6|\hat{\gamma}|r_3, r_4\rangle\langle r_3, r_4|\hat{\Phi}|r'_1, r'_2\rangle.$$

Therefore, we can find the real-space representation of the BS as

$$\Phi(r_1, r'_1; r_2, r'_2) = G^R(r_1, r'_1) G^A(r_2, r'_2) \qquad (5.25)$$
$$+ \sum_{r_3, r_4, r_5, r_6} G^R(r_1, r_5) G^A(r_2, r_6) \gamma(r_5, r_3; r_6, r_4) \Phi(r_3, r'_1; r_4, r'_2).$$

The summation over the real-space basis is in this context equivalent to integration, because the the real-space is continuous

To derive and further analyse the kinetic equation, which is equivalent to the Boltzmann equation, we return to the BS in operator notation

$$\hat{\Phi} = \left(\hat{\mathbf{G}}^R \otimes \hat{\mathbf{G}}^A \right) \left[\mathbb{1} \otimes \mathbb{1} + \hat{\gamma}\hat{\Phi} \right]. \qquad (5.26)$$

Now we use the operator identity valid for any operators $\hat{A}$ and $\hat{B}$ which has also been employed in chapter three

$$\hat{A} \otimes \hat{B} = (\hat{A}^{-1} \otimes \mathbb{1} - \mathbb{1} \otimes \hat{B}^{-1})^{-1} [\mathbb{1} \otimes \hat{B} - \hat{A} \otimes \mathbb{1}] \qquad (5.27)$$

to the term $\left(\hat{\mathbf{G}}^R \otimes \hat{\mathbf{G}}^A \right)$ and arrive at the kinetic equation in operator notation

$$\left[(\hat{\mathbf{G}}^R)^{-1} \otimes \mathbb{1} - \mathbb{1} \otimes (\hat{\mathbf{G}}^A)^{-1} \right] \hat{\Phi} \qquad (5.28)$$
$$= \left(\mathbb{1} \otimes \hat{\mathbf{G}}^A - \hat{\mathbf{G}}^R \otimes \mathbb{1} \right) \left[\mathbb{1} \otimes \mathbb{1} + \hat{\gamma}\hat{\Phi} \right].$$

The real-space representation is found in the same way as the above BS equation itself, i.e. we multiply with $\langle r_1, r_2|$ from left and with $|r'_1, r'_2\rangle$ from right and insert unit operators $\mathbb{1} = |r_7, r_8\rangle\langle r_7, r_8|$ at appropriate places, and arrive at

$$\sum_{r_7,r_8} \langle r_1, r_2| \left[(\hat{\mathbf{G}}^R)^{-1} \otimes \mathbb{1} - \mathbb{1} \otimes (\hat{\mathbf{G}}^A)^{-1} \right] |r_7, r_8\rangle\langle r_7, r_8|\hat{\boldsymbol{\Phi}}|r'_1, r'_2\rangle$$
$$= \langle r_1, r_2| \left(\mathbb{1} \otimes \hat{\mathbf{G}}^A - \hat{\mathbf{G}}^R \otimes \mathbb{1} \right) |r'_1, r'_2\rangle \qquad (5.29)$$
$$+ \sum_{r_3,r_4,r_5,r_6} \langle r_1, r_2| \left(\mathbb{1} \otimes \hat{\mathbf{G}}^A - \hat{\mathbf{G}}^R \otimes \mathbb{1} \right) |r_5, r_6\rangle\langle r_5, r_6|\hat{\gamma}|r_3, r_4\rangle\langle r_3, r_4|\hat{\boldsymbol{\Phi}}|r'_1, r'_2\rangle.$$

The operator $\hat{\mathbf{G}}^R$ was by definition the inverse differential operator of the defining differential equation (wave equation), therefore $(\hat{\mathbf{G}}^R)^{-1}$ represents the original resolvant operator for the wave equation. Bearing this in mind one finds for the action of $\hat{\mathbf{G}}^R$ on the combined Ket-vector $|r_7, r_8\rangle$ in Eq. (5.29)

$$\langle r_1, r_2|(\hat{\mathbf{G}}^R)^{-1} \otimes \mathbb{1}|r_7, r_8\rangle = \langle r_1|(\hat{\mathbf{G}}^R)^{-1}|r_7\rangle \otimes \langle r_2|r_8\rangle$$
$$= \left(\epsilon_b \omega_+^2 + \nabla_{r_1}^2 - \Sigma^R \right) \mathbb{1} \delta_{r_1,r_7} \otimes \mathbb{1}\delta_{r_2,r_8} \qquad (5.30)$$

where ∇_{r_1} implies differentiation with respect to r_1. The placement of the unit operator $\mathbb{1}$ is necessary to clarify the matching of the spaces the operators live in. In complete analogy we also find for the advanced Green's function

$$\langle r_1, r_2|\mathbb{1} \otimes (\hat{\mathbf{G}}^A)^{-1}|r_7, r_8\rangle = \langle r_1|r_7\rangle \otimes \langle r_2|(\hat{\mathbf{G}}^A)^{-1}|r_8\rangle$$
$$= \delta_{r_1,r_7} \mathbb{1} \otimes \left(\epsilon_b^* \omega_-^2 + \nabla_{r_2}^2 - \Sigma^A \right) \mathbb{1}\delta_{r_2,r_8}. \qquad (5.31)$$

With this the above equation, Eq. (5.29), can be written as

$$[(\Sigma^A - \Sigma^R) + (\epsilon_b \omega_+^2 - \epsilon_b^* \omega_-^2) + (\nabla_{r_1}^2 - \nabla_{r_2}^2)] \Phi(r_1, r'_1; r_2, r'_2)$$
$$= [G^A(r_2, r'_2) - G^R(r_1, r'_1)] \qquad (5.32)$$
$$+ \sum_{r_3,r_4,r_5,r_6} [G^A(r_2, r_6) - G^R(r_1, r_5)]$$
$$\times \gamma(r_5, r_3; r_6, r_4)\Phi(r_3, r'_1; r_4, r'_2).$$

This kinetic equation, Eq. (5.32), is as an integro-differential equation analogous to the so-called Boltzmann equation in real-space representation.
As we have discussed above, it is necessary to employ a partial Fourier-transformation. For this purpose we have to transform to center of mass and relative coordinates according to the transformation rules shown in Appendix E. The kinetic equation in real-space representation, Eq. (5.32) reads now

$$[(\Sigma^A - \Sigma^R) + (\epsilon_b \omega_+^2 - \epsilon_b^* \omega_-^2) + (2\nabla_R \nabla_{\Delta r})] \qquad (5.33)$$
$$\times \Phi(R + \frac{\Delta r}{2}, R' + \frac{\Delta r'}{2}; R - \frac{\Delta r}{2}, R' - \frac{\Delta r'}{2})$$
$$= \left[G^A(R - \frac{\Delta r}{2}, R' - \frac{\Delta r'}{2}) - G^R(R + \frac{\Delta r}{2}, R' + \frac{\Delta r'}{2}) \right]$$
$$+ \sum_{R_{34},R_{56},\Delta r_{34},\Delta r_{56}} \left[G^A(R - \frac{\Delta r}{2}, R_{56} - \frac{\Delta r_{56}}{2}) - G^R(R + \frac{\Delta r}{2}, R_{56} + \frac{\Delta r_{56}}{2}) \right]$$
$$\times \gamma(R_{56} + \frac{\Delta r_{56}}{2}, R_{34} + \frac{\Delta r_{34}}{2}; R_{56} - \frac{\Delta r_{56}}{2}, R_{34} - \frac{\Delta r_{34}}{2})$$
$$\times \Phi(R_{34} + \frac{\Delta r_{34}}{2}, R' + \frac{\Delta r'}{2}; R_{34} - \frac{\Delta r_{34}}{2}, R' - \frac{\Delta r'}{2}).$$

The single particle Green's function depends only on the relative coordinates as shown section 5.1. Therefore (5.33) is written as follows

$$[(\Sigma^A - \Sigma^R) - \epsilon_b \omega_+^2 - \epsilon_b^* \omega_-^2) + (2\nabla_R \nabla_{\Delta r})]$$
$$\times \Phi(R + \frac{\Delta r}{2}, R' + \frac{\Delta r'}{2}; R - \frac{\Delta r}{2}, R' - \frac{\Delta r'}{2})$$
$$= [G^A(\Delta r) - G^R(\Delta r)] \delta(\Delta r - \Delta r')$$
$$+ \sum_{R_{34}, \Delta r_2} \left[G^A(R - \frac{\Delta r}{2}) - G^R(R + \frac{\Delta r}{2}) \right] \delta(\Delta r - \Delta r_{56})$$
$$\times \gamma(R + \frac{\Delta r}{2}, R_{34} + \frac{\Delta r_{34}}{2}; R - \frac{\Delta r}{2}, R_{34} - \frac{\Delta r_{34}}{2})$$
$$\times \Phi(R_{34} - \frac{\Delta r_{34}}{2}, R' + \frac{\Delta r'}{2}; R_{34} - \frac{\Delta r_{34}}{2}, R' - \frac{\Delta r'}{2}) \quad (5.34)$$

5.3 Fourier - Transformed Bethe - Salpeter Equation

We begin with the above derived Bethe - Salpeter equation, Eq.(5.34). Due to physical reasons, all quantities are translational invariant with respect to the relative spatial coordinate, therefore we introduce a Fourier transformation according to

$$G_p = \int d\Delta r \, e^{-ip\cdot\Delta r} G(\Delta r) \quad (5.35)$$

$$G(\Delta r) = \int \frac{dp}{(2\pi)^3} e^{+ip\cdot\Delta r} G_p. \quad (5.36)$$

Since the single-particle Green's function depends only on the spatial relative coordinate (translational invariance) it follows that it depends on one momentum only. Therefore we have

$$G_{pp'} = G_p \delta(p - p'). \quad (5.37)$$

After applying the in-plane Fourier-transformation with respect to the center of mass coordinate in Eq. (5.34), the delta function $\delta(Q_\| - Q'_\|)$ appears in all terms and may therefore be omitted from now on, furthermore we introduce the shorthands ΔX and $\Box X$, for any quantity X according to

$$\Delta \Sigma \equiv \Sigma^A - \Sigma^R \quad (5.38)$$
$$\Box \Sigma \equiv \Sigma^A + \Sigma^R \quad (5.39)$$

and in general for any quantity as for instance the Green's function

$$\Delta G(Q, \Omega) \equiv \left(G_{q-}^{\omega-}\right)^A - \left(G_{q+}^{\omega+}\right)^R \quad (5.40)$$

$$\Box G(Q, \Omega) \equiv \left(G_{q-}^{\omega-}\right)^A + \left(G_{q+}^{\omega+}\right)^R \quad (5.41)$$

and can finally rewrite the above kinetic equation Eq. (5.34) as

$$\left[\Delta \Sigma + \Box \epsilon_b \omega \Omega - \Delta \epsilon_b \omega^2 - 2\vec{p}_\| \cdot \vec{Q}_\| + 2ip_z \partial_Z \right] \Phi_{pp'}^{Q_\|}(Z, Z')$$
$$= \Delta G_p(Q_\|) \delta(p - p')$$
$$+ \sum_{Z_{34}} \Delta G_p(Q_\|) \int \frac{dp''}{(2\pi)^3} \gamma_{pp''}^{Q_\|}(Z, Z_{34}) \Phi_{p''p'}^{Q_\|}(Z_{34}, Z') \quad (5.42)$$

The single terms in the above equation have a physical interpretation. $\Delta\Sigma$ represents single-particle scattering of the intensity, the term proportional to Ω represents the temporary change of the intensity since Ω is the Fourier-partner of the time. The last two terms $2\vec{p}_{\parallel}\cdot\vec{Q}_{\parallel}$ and $2ip_z\partial_Z$ represent the drift terms, because their Fourier-transform is equal to the spatial derivative. The last term on the right hand side is the so called collision integral which includes here also interference effects.

5.4 Light Intensity Transport in Bounded Disordered Media with Absorption or Gain

In this chapter we consider the solution of the above derived Bethe-Salpeter equation, Eq. (5.25) for systems in a film-geometry, which we state here again in its form prior to the Fourier transformation of the relative coordinate Δz along the z-direction

$$\begin{aligned}\Phi(r_1,r_1';r_2,r_2') &= G^R(r_1,r_1')G^A(r_2,r_2') \\ &+ \sum_{r_3,r_4,r_5,r_6} G^R(r_1,r_5)G^A(r_2,r_6)\gamma(r_5,r_3;r_6,r_4)\Phi(r_3,r_1';r_4,r_2').\end{aligned} \quad (5.43)$$

Introducing relative and center of mass coordinates and further utilizing the operator notation, as discussed in detail above, we may rewrite the Bethe-Salpeter equation.
Eventually, we apply the following operator identity

$$\hat{A}\hat{B} = \frac{\hat{B}-\hat{A}}{\hat{A}^{-1}-\hat{B}^{-1}} \quad (5.44)$$

in order to derive the *Boltzmann equation* of transport, also referred to as the *kinetic equation*, and given by

$$\begin{aligned}\left[\Delta\Sigma + \Box\epsilon\omega\Omega - \Delta\epsilon\omega^2 - 2\vec{p}_{\parallel}\cdot\vec{Q}_{\parallel} + 2ip_z\partial_Z\right]\Phi^{Q_{\parallel}}_{pp'}(Z,Z') \\ = \Delta G_P(Q_{\parallel};Z,Z')\delta(p-p') \\ + \sum_{Z_{34}}\Delta G_p(Q_{\parallel})\int\frac{\mathrm{d}p''}{(2\pi)^3}\gamma^{Q_{\parallel}}_{pp''}(Z,Z_{34})\Phi^{Q_{\parallel}}_{p''p'}(Z_{34},Z')\end{aligned} \quad (5.45)$$

Details of the derivation can be found in section 5.2.

5.4.1 Solution by Moment Expansion

We follow the general solution procedure already discussed in the context of light propagation in infinite three dimensional media. This means we use an expansion of the intensity correlator Φ which is to be calculated into its moments. The first two terms of expansion are identified as the energy density (the zeroth moment) and the energy density current (the first moment) and subsequently related to each other by the derivation of two independent equations.
To obtain the continuity equation we first define the energy density $\Phi_{\epsilon\epsilon}$ and energy density current $\Phi_{j\epsilon}$ in terms of $\Phi^{\omega}_{p'p}(\vec{Q},\Omega;Z,Z')$ and also in terms of $\Phi_{\rho\rho}$ and $\Phi_{j\rho}$ (the intrinsic first two moments of the correlator Φ) according to

$$\Phi_{\epsilon\epsilon} = \frac{\omega^2}{c_{ph}^2}\Phi_{\rho\rho} = \frac{\omega^2}{c_{ph}^2}\sum_{p,p'}\Phi^{\omega}_{p'p}(\vec{Q},\Omega;Z,Z') \quad (5.46)$$

$$\Phi_{j\epsilon} = \frac{\omega}{c_{ph}}v_E(\omega)\Phi_{j\rho} = \frac{\omega}{c_{ph}}v_E(\omega)\sum_{p,p'}\left(2\vec{k}\cdot\hat{Q}\right)\Phi^{\omega}_{p'p}(\vec{Q},\Omega;Z,Z'). \quad (5.47)$$

5.4. LIGHT INTENSITY TRANSPORT IN BOUNDED DISORDERED MEDIA WITH ABSORPTION OR GAIN

In the above equation, Eq (5.47) we have used a three dimensional vector $\hat{Q}$, which requires some explanation. We start with the definition of a three dimensional vector operator $\widetilde{\mathbf{Q}}$,

$$\widetilde{\mathbf{Q}} = \begin{pmatrix} Q_x \\ Q_y \\ -i\partial_Z \end{pmatrix} \qquad (5.48)$$

whose action on a given testfunction $\varphi(Q_x, Q_y, Z)$ is defined according to

$$\begin{aligned}\widetilde{\mathbf{Q}}^2 \varphi(Q_x, Q_y, Z) &\equiv (Q_x^2 + Q_y^2 - \partial_Z^2)\varphi(Q_x, Q_y, Z) \\ &= Q_x^2 \varphi(Q_x, Q_y, Z) + Q_y^2 \varphi(Q_x, Q_y, Z) - \partial_Z^2 \varphi(Q_x, Q_y, Z).\end{aligned} \qquad (5.49)$$

$\hat{Q}$ points to the direction of $\widetilde{\mathbf{Q}}$.

5.4.2 Continuity Equation

The first step consists in deriving a first equation relating energy density and energy density current. The result of this will be the *continuity equation*. It is obtained by summing Eq. (5.45) over the (three-dimensional) vectors $\vec{p}$ and also over $\vec{p}'$

$$\begin{aligned}&\sum_{pp'} \Delta\Sigma(\omega)\Phi^\omega_{pp'}(\vec{Q},\Omega;Z,Z') - \Delta\epsilon\omega^2 \Phi_{\rho\rho}(\vec{Q},\Omega;Z,Z') \\ &+ \Box\epsilon\omega^2\Phi_{\rho\rho}(\vec{Q},\Omega;Z,Z') - Q\Phi_{j\rho}(\vec{Q},\Omega;Z,Z') \\ &= \Delta G_0 + \sum_{pp',Z_{34}} \Delta G_p(\vec{Q},\Omega)\gamma^\omega_{pp''}(\vec{Q},\Omega;Z,Z_{34})\Phi^\omega_{p''p'}(\vec{Q},\Omega;Z_{34},Z').\end{aligned} \qquad (5.50)$$

where we used the definitions of the moments as given in Eq. (5.46) and Eq. (5.47).

Again, we incorporate the local energy conservation by means of the Ward identity, which we also calculated for this particular geometry, i.e. for the case of spatial dependence on the center of mass coordinates (Z,Z'), *c.f.* Appendix C. This Ward identity employed by substituting the first term on the left hand side, $\sum_p \Delta\Sigma(\omega)\Phi^\omega_{pp'}(\vec{Q},\Omega;Z,Z')$, in the above expression, Eq. (5.50).
For the sake of completeness we state the Ward identity here, which reads for these systems of a film of finite width d

$$\begin{aligned}\Delta\Sigma^\omega_p(\vec{Q},\Omega) &= \sum_{p'}\sum_{Z''}\Delta G_{p'}(\vec{Q},\Omega)\gamma_{p'p}(\vec{Q},\Omega;Z'',Z') \\ &+ f_\omega(\Omega)\Big[\sum_{p'}\sum_{Z''}\Box G_{p'}(\vec{Q},\Omega)\gamma_{p'p}(\vec{Q},\Omega;Z'',Z') + \Box\Sigma^\omega_p(\vec{Q},\Omega)\Big]\end{aligned} \qquad (5.51)$$

where we introduce the abbreviation as above $\Box\Sigma_p(\omega) \equiv \Sigma^R_{p_+}(\omega) + \Sigma^A_{p_-}(\omega)$ etc., and the renormalization factor $f(\omega)$ is given by

$$f_\omega(\Omega) = \frac{(\omega\Omega\mathrm{Re}\,\Delta\epsilon + i\omega^2\mathrm{Im}\,\Delta\epsilon)}{(\omega^2\mathrm{Re}\,\Delta\epsilon + i\omega\Omega\mathrm{Im}\,\Delta\epsilon)}. \qquad (5.52)$$

Using the Ward identity to substitute $\Delta\Sigma$ in Eq. (5.50) and afterwards employing an expansion for $(Q,\Omega) \to (0,0)$ on the substituted terms we first observe a cancellation with respect to the last term on the r.h.s. of Eq. (5.50), and finally using the relations given in Eq. (5.46) and in Eq. (5.47) we eventually arrive at

$$\begin{aligned}\Omega\Phi_{\epsilon\epsilon}(\vec{Q},\Omega)g^{(1)}_\omega\Big(1+\Delta(\omega)\Big) &- \frac{\omega}{c_p v_E}Q\Phi_{j\epsilon}(\vec{Q},\Omega) \\ &= \frac{\omega^2}{c_p^2}\Delta G_0(\vec{Q},\Omega) + \Phi_{\epsilon\epsilon}(\vec{Q},\Omega)i\Lambda(\omega)\end{aligned} \qquad (5.53)$$

or equivalently written in the following way

$$\Omega \Phi_{\epsilon\epsilon}(\vec{Q},\Omega) - Q\Phi_{j\epsilon}(\vec{Q},\Omega) \tag{5.54}$$
$$= \frac{\omega^2 \Delta G_0(\vec{Q},\Omega)}{c_p^2 g_\omega^{(1)}\left(1+\Delta(\omega)\right)} + \Phi_{\epsilon\epsilon}(\vec{Q},\Omega)\frac{i\Lambda(\omega)}{g_\omega^{(1)}\left(1+\Delta(\omega)\right)}.$$

In the above equations, especially in the continuity equation, Eq. (5.54), we introduced the following abbreviations

$$\Delta G_0 \equiv \int \frac{d^3 p}{(2\pi)^3} \Delta G_p^\omega(Q,\Omega) \qquad \Delta\epsilon = \epsilon_{scat} - \epsilon_b \tag{5.55}$$

$$u_\epsilon = \frac{\operatorname{Im}(\Delta\epsilon\Sigma)}{\operatorname{Im}(\Delta\epsilon G_0(\omega))} \qquad r_\epsilon = \frac{\operatorname{Im}\Delta\epsilon}{\operatorname{Re}\Delta\epsilon} \tag{5.56}$$

$$A_\epsilon = 2\left[u_\epsilon \operatorname{Re} G_0 + \operatorname{Re}\Sigma\right] \qquad B_\epsilon = \frac{(\operatorname{Re}\Delta\epsilon)^2 + (\operatorname{Im}\Delta\epsilon)^2}{2\omega^2 (\operatorname{Re}\Delta\epsilon)^2} \tag{5.57}$$

as well as

$$v_{\mathrm{E}} = \frac{c^2}{c_p \operatorname{Re}\epsilon_b}\frac{1}{1+\Delta(\omega)} \tag{5.58}$$

$$\Delta(\omega) = B_\epsilon A_\epsilon + i r_\epsilon \partial_\Omega A_\epsilon(\Omega) \tag{5.59}$$

$$\Lambda(\omega) = i\omega^2 \operatorname{Im}\epsilon_b - i r_\epsilon A_\epsilon$$

$$g_\omega^{(0)} = \frac{2\omega}{c^2}\operatorname{Im}\epsilon_b \qquad g_\omega^{(1)} = \frac{4\omega}{c^2}\operatorname{Re}\epsilon_b \tag{5.60}$$

$$\tilde{A} = \frac{c_p v_{\mathrm{E}}}{\omega}\left[1 - \frac{\omega^2 \operatorname{Im}\epsilon_b}{u_\epsilon \operatorname{Im} G_0} + \frac{r_\epsilon A_\epsilon}{u_\epsilon \operatorname{Im} G_0}\right]. \tag{5.61}$$

The continuity equation, Eq. (5.53), relates the temporal derivative of the energy density (here Fourier transformed to Ω) to the divergence of the energy density current (here Fourier transformed to Q). The sum of both equals the source (drain) of energy density in the system, therefore it represents in this sense energy conservation.

In the above equation, Eq. (5.53), we recover the correction terms, which we also have found for three dimensional infinite media. This had to be expected, since they originate from the Ward identity and therefore represent the non-conserving characteristics of the systems.

Despite the recovering of previously found corrections, Eq. (5.54) is severely different from the continuity equation for infinite media in that it contains a differential operator, and therefore establishes a differential equation.

5.4.3 Current Relaxation Equation

As explained in chapter one and discussed in some detail in chapter three, we need a second and linearly independent equation, which relates the energy density to the energy density current. This relation is called current density relation.

We obtain this relation, by going back to the Boltzmann equation, Eq. (5.45), and multiply it with the current projector $\vec{p}\cdot\hat{Q}$. Afterwards we sum the equation over p and p'. This procedure yields

$$\left[\Delta\Sigma(\omega) - \Delta\epsilon\omega^2 + \Box\epsilon\omega\Omega\right]\Phi_{j\rho} - \tilde{Q}\sum_p \left(2\vec{p}\cdot\hat{Q}\right)^2 \Phi_p^\omega(\vec{Q},\Omega) \tag{5.62}$$

$$= \sum_p \left(2\vec{p}\cdot\hat{Q}\right)\Delta G_p + \sum_{pp''}\left(2\vec{p}\cdot\hat{Q}\right)\Delta G_p \gamma_{pp''}^\omega(\vec{Q},\Omega;Z)\Phi_{p''p'}^\omega(\vec{Q},\Omega).$$

5.5. DIFFUSION POLE STRUCTURE

At this point we use the moment expansion of the energy density correlator as given by

$$\Phi_{\overline{p}} \approx \frac{\Delta G_p}{\Delta G_0}\Phi_{\rho\rho} + \frac{\left(2\vec{p}\cdot\hat{Q}\right)\Delta G_p^2}{\sum_k \left(2\vec{k}\cdot\hat{Q}\right)\Delta G_k^2}\Phi_{j\rho} + \ldots \qquad (5.63)$$

as well as the Ward identity as explained above and afterwards we again expand the expression for $(Q,\Omega) \to (0,0)$. The result of this procedure is eventually

$$\left[\text{Re}\epsilon_b\omega\Omega - i\left[\text{Im}\Sigma(\omega) - \text{Im}\Delta\epsilon\omega^2\right] + iM(\Omega)\right]\Phi_{j\epsilon} + \widetilde{Q}\Phi_{\epsilon\epsilon}\tilde{A} = 0 \qquad (5.64)$$

where we introduced the so-called memory kernel $M(\Omega)$ by the following definition

$$\begin{aligned}-iM(\Omega) &= \frac{1}{\int \frac{dk}{(2\pi)^2}(2\vec{k}\cdot\hat{Q})(\Delta G_k)^2}\int\frac{dk}{(2\pi)^2}\int\frac{dk'}{(2\pi)^2} \\ &\quad \times (2\vec{k}\cdot\hat{Q})(\Delta G_k)\gamma_{k\,k'}(\Delta G_{k'})^2(2\vec{k}'\cdot\hat{Q})\end{aligned} \qquad (5.65)$$

The above derived current density relation or current relaxation equation, Eq. (5.64) is also a differential equation, since it contains the vector operator $\widetilde{Q}$. The calculation of the memory kernel M is explicitly shown in Appendix D.

5.5 Diffusion Pole Structure

In the previous subsections we have detailed the solution scheme of the Bethe-Salpeter-equation, Eq. (5.43), with the help of a moment expansion of the intensity correlator Φ. As presented in detail, this yields two linearly independent equation relating the two quantities energy density and energy density current. Consequentially, Eq. (5.54) and Eq. (5.64) constitute an algebraic relation between energy density and energy density current which now includes contributions proportional two the second derivatives of those quantities.
We therefore apply the operator $\widetilde{Q}$ on Eq. (5.64) and use the result to substitute $\widetilde{Q}\Phi_{j\rho}$ in Eq. (5.53). As a result we obtain

$$\widetilde{Q}\left[\Box\epsilon\omega + \Delta\Sigma(\omega) - \Delta\epsilon\omega^2 + iM(\omega)\right]\Phi_{j\rho} + \widetilde{Q}^2\Phi_{\rho\rho}W = 0 \qquad (5.66)$$

$$\widetilde{Q}\left[\text{Re}\epsilon_b\omega\Omega + i\left[\text{Im}\Sigma(\omega) - \text{Im}\Delta\epsilon\omega^2\right] + iM(\Omega)\right]\Phi_{j\epsilon} + \widetilde{Q}^2\Phi_{\epsilon\epsilon}\tilde{A} = 0. \qquad (5.67)$$

For the sake of a clear notation let us define the abbreviation [1] according to

$$[1] := \left[\text{Re}\epsilon_b\omega\Omega + i\left[\text{Im}\Sigma(\omega) - \text{Im}\Delta\epsilon\omega^2\right] + iM(\Omega)\right]. \qquad (5.68)$$

Recalling the definition of the operator $\widetilde{Q}^2$

$$\widetilde{Q}^2 = Q_x^2 + Q_y^2 - \partial_Z^2 \qquad (5.69)$$

we define

$$Q_\parallel^2 := Q_x^2 + Q_y^2 \qquad (5.70)$$

and find e.g.

$$\frac{\tilde{A}\left(\tilde{Q}^2\Phi_{\epsilon\epsilon}\right)}{[1]\Phi_{\epsilon\epsilon}} = Q_{\parallel}^2 \frac{\tilde{A}}{[1]} - \frac{\left(\partial_Z^2\Phi_{\epsilon\epsilon}\right)\tilde{A}}{[1]\Phi_{\epsilon\epsilon}} \tag{5.71}$$

and also

$$\frac{\left(\tilde{Q}^2\tilde{A}\right)}{[1]} = Q_{\parallel}^2 \frac{\tilde{A}}{[1]} - \frac{\left(\partial_Z^2\tilde{A}\right)}{[1]} = Q_{\parallel}^2 \frac{\tilde{A}}{[1]} \tag{5.72}$$

since the quantity $\tilde{A}$ consists according to its definition Eq. (5.61) of single-particle quantities only, which do not show a functional dependence on Z. Hence the derivatives vanish. Using this together with the identity $\left(\tilde{Q}^2\Phi_{\epsilon\epsilon}\right) \equiv \Phi_{\epsilon\epsilon}\frac{(\tilde{Q}^2\Phi_{\epsilon\epsilon})}{\Phi_{\epsilon\epsilon}}$ we may rewrite the above equation, Eq. (5.67), in the following formal way

$$\left[\Omega + \frac{\left(\tilde{Q}^2\Phi_{\epsilon\epsilon}\right)\tilde{A}}{[1]\Phi_{\epsilon\epsilon}} + \frac{\Phi_{j\epsilon}\left(\tilde{Q}[1]\right)}{[1]\Phi_{\epsilon\epsilon}} \right. \tag{5.73}$$

$$\left. + \frac{2\left(\tilde{Q}\Phi_{\epsilon\epsilon}\right)\left(\tilde{Q}\tilde{A}\right)}{[1]\Phi_{\epsilon\epsilon}} + \frac{\left(\tilde{Q}^2\tilde{A}\right)}{[1]} - \frac{i\Lambda(\omega)}{g_\omega^{(1)}\left(1+\Delta(\omega)\right)}\right]\Phi_{\epsilon\epsilon}$$

$$= \frac{\omega^2 \Delta G_0(\vec{Q},\Omega)}{c_p^2 g_\omega^{(1)}\left(1+\Delta(\omega)\right)}. \tag{5.74}$$

By introducing the definitions of the numerator of the diffusion pole $N_\omega(Z)$ as

$$N_\omega(Z) := \frac{\omega^2 \Delta G_0(\vec{Q},\Omega)}{c_p^2 g_\omega^{(1)}\left(1+\Delta(\omega)\right)} \tag{5.75}$$

and the diffusion coefficient $D(\Omega, Z)$ for the slab geometry

$$-iD(\Omega,Z) := \left\{\frac{2\tilde{A}}{[\text{Re}\epsilon_b\omega\Omega + i\left[\text{Im}\Sigma(\omega) - \text{Im}\Delta\epsilon\omega^2\right] + iM(\Omega)]}\right\} \tag{5.76}$$

as well as the dissipation length scale $\chi_d(Z)$, which is describing the growth or the absorption of intensity and which is also emerging in the infinite system,

$$\frac{iD(\Omega,Z)}{\chi_d(Z)^2} := \frac{i\Lambda(\omega)}{g_\omega^{(1)}\left(1+\Delta(\omega)\right)} \tag{5.77}$$

we can formally extract the pole structure as

$$\Phi_{\epsilon\epsilon}(Q,\Omega) = \frac{N_\omega(Z)}{\Omega + iDQ_{\parallel}^2 - iD\chi_d^{-2} - \frac{(\partial_Z^2\Phi_{\epsilon\epsilon})\tilde{A}}{[1]\Phi_{\epsilon\epsilon}} + \frac{2(\tilde{Q}\Phi_{\epsilon\epsilon})(\tilde{Q}\tilde{A})+\Phi_{j\epsilon}(\tilde{Q}[1])}{[1]\Phi_{\epsilon\epsilon}}}. \tag{5.78}$$

To gain a better insight we want to stress again that the phenomenon of diffusion relies on conservation laws and may be analyzed by means of the presence of a diffusion pole in the energy density. The general idea behind this was explained in chapter 2, and it is based on the fact that in leading order diffusion is described linearly in Ω and quadratically in Q. Corrections to this behavior are included in an Ω-dependent diffusion constant, which is determined selfconsistently by including Cooperon contributions. This was done in the pioneering work by D. Vollhardt and P. Wölfle [65]. Here we are interested in studying the wave transport phenomena in dissipative media of finite width due to geometrical restrictions. However, for comparability and consistency reasons we also

5.5. DIFFUSION POLE STRUCTURE

apply a hydrodynamic expansion, i.e. $(Q,\Omega) \rightarrow (0,0)$, to extract the diffusion pole structure and study diffusion in the presence of self-interference of the waves by means of a frequency dependent diffusion coefficients.
This in turn has consequences for the diffusion pole shown in Eq. (5.78) regarding the term containing the second derivative of $\Phi_{\epsilon\epsilon}$. It is to be taken at $Q_x = 0 = Q_y$ and $\Omega = 0$, since, as just explained, this contains and reveals the diffusive characteristics of the system. In order to extract the correlation length we have to find an expression for $-\frac{(\partial_Z^2 \Phi_{\epsilon\epsilon})\tilde{A}}{[1]\Phi_{\epsilon\epsilon}}$ at $Q_x = 0 = Q_y$ and at $\Omega = 0$.
At this point $(Q_x = 0 = Q_y, \Omega = 0)$ the equation determining $\Phi_{\epsilon\epsilon}$, Eq. (5.73), reads

$$\frac{\left(-\frac{\partial^2}{\partial Z^2}\Phi_{\epsilon\epsilon}\right)\tilde{A}}{[1]\Phi_{\epsilon\epsilon}} = \frac{\omega^2 \Delta G_0(\vec{Q},\Omega)}{c_p^2 g_\omega^{(1)}\left(1+\Delta(\omega)\right)\Phi_{\epsilon\epsilon}} + \frac{i\Lambda(\omega)}{g_\omega^{(1)}\left(1+\Delta(\omega)\right)}. \tag{5.79}$$

Using this expression (and recognizing the last term to be D/χ_d^2) to substitute it in the diffusion pole we obtain for the energy density

$$\Phi_{\epsilon\epsilon}(Q,\Omega) = \frac{N_\omega(Z)}{\Omega + iDQ_\parallel^2 + iD\cdot\xi^{-2}} \tag{5.80}$$

which defines the positive correlation length ξ as

$$\frac{1}{\xi^2} = \frac{N_\omega(Z)}{D\Phi_{\epsilon\epsilon}(Q_\parallel = 0, \Omega = 0)} \tag{5.81}$$

Eventually the correlation length ξ, Eq. (5.81), the diffusion coefficient $D(\Omega)$, Eq. (5.76), and the energy density $\Phi_{\epsilon\epsilon}$, Eq. (5.80), constitute the solution of the transport theory of light intensity within a finite system with boundaries.

Selfconsistent System of Equations

Here we state again the main equations from the above derivation, to have them present in one place. The equations that constitute the solution are:
The Energy-Density

$$\Phi_{\epsilon\epsilon}(Q,\Omega) = \frac{N_\omega(Z)}{\Omega + iDQ_\parallel^2 + iD\cdot\xi^{-2}} \tag{5.82}$$

the Diffusion-Constant

$$D(\Omega)\left[1 - i\Omega\tau_a^2 \text{Re}\epsilon\omega\right] = D_0^{tot} - \tau_a^2 D(\Omega)M(\omega) \tag{5.83}$$

the Correlation-Length

$$\frac{1}{\xi^2} = \frac{N_\omega(Z)}{D\Phi_{\epsilon\epsilon}(Q_\parallel = 0, \Omega = 0)} \tag{5.84}$$

the differential equation for the Energy-Density

$$-\frac{\partial^2}{\partial Z^2}\Phi_{\epsilon\epsilon} = \frac{\omega^2 \Delta G_0(\vec{Q},\Omega)[1]}{c_p^2 g_\omega^{(1)}\left(1+\Delta(\omega)\right)\tilde{A}} + \frac{i\Lambda(\omega)[1]}{g_\omega^{(1)}\left(1+\Delta(\omega)\right)\tilde{A}}\Phi_{\epsilon\epsilon} \tag{5.85}$$

Now recognizing that in the above equation, Eq. (5.85), the last term is just $-\frac{1}{\chi_d^2}\Phi_{\epsilon\epsilon}$ with χ_d the grow length as also found in infinite media and defined in Eq. (5.77) and defined above, we rewrite Eq. (5.85) as

$$-\frac{\partial^2}{\partial Z^2}\Phi_{\epsilon\epsilon} = \frac{1}{D}\left[\frac{D}{-\chi_d^2}\right]\Phi_{\epsilon\epsilon} + \frac{1}{D}\frac{\omega^2 \Delta G_0(\vec{Q},\Omega)}{c_p^2 g_\omega^{(1)}\left[1+\Delta(\omega)\right]}. \tag{5.86}$$

5.5.1 Selfconsistent Diffusion Coefficient

In this section we derive the selfconsistent equation, which determines the diffusion coefficient $D(\Omega, Z)$ defined Eq. (5.76), together with the definitions Eqs. (5.55-5.61), as well as Eq. (5.65) we finally find

$$D(\Omega)\left[1 - i\Omega \mathrm{Re}\epsilon\omega\tau_a^2\right] = D_0^{tot} - \tau_a^2 D(\Omega) M(\omega) \tag{5.87}$$

where we have used

$$\Delta\Sigma(\omega) \equiv 2i\mathrm{Im}\Sigma(\omega) \tag{5.88}$$
$$\Box\epsilon_b \equiv 2\mathrm{Re}\epsilon_b. \tag{5.89}$$

Furthermore we introduced the lifetime τ_a of the diffusing modes, due to both, scattering and non-conservation of energy, as

$$\tau_a^2 = \frac{1}{\mathrm{Im}\Sigma - \mathrm{Im}\epsilon\omega^2} \tag{5.90}$$

and the diffusion constant without memory effects, $D_0^{tot} = D_0 + D_b + D_s$, consisting of the bare diffusion constant,

$$D_0 = \frac{2v_E c_p}{\omega \Delta G} \int \frac{d^2 k}{(2\pi)^2} \, [\vec{k}\cdot\hat{Q}]^2 (\Delta G_{\vec{k}}^\omega)^2 \tag{5.91}$$

and dissipative contributions (renormalizations) from absorption or gain in the background medium (D_b) and in the scatterers (D_s), given by

$$D_b = \frac{1}{4}(\omega\tau_a)^2 \, \Delta\epsilon_b \, \tilde{D}_0 \tag{5.92}$$
$$D_s = \frac{1}{8}r_\epsilon A_\epsilon \tau_a^2 \tilde{D}_0. \tag{5.93}$$

These two contributions have already been found in chapter three for infinite media. They originate from the Ward-Identity and therefore have only a small parametric dependency. Eq. (5.87) establishes a self-consistency relation for D since the memory function M itself depends on the diffusion coefficient since the irreducible vertex function γ contained in M is expressed by the diffusion pole of the energy density correlation function as shown in Appendix D, and therefore depends on the diffusion coefficient D.

5.6 Microscopic Theory of Random Lasers

To derive a full solution for the Random Laser, we have to couple the transport theory of light to the laser rate equations. As discussed in detail in chapter four, the rate equations for a four-level laser are

$$\frac{\partial N_3}{\partial t} = \frac{N_0}{\tau_P} - \frac{N_3}{\tau_{32}} \tag{5.94}$$

$$\frac{\partial N_2}{\partial t} = \frac{N_3}{\tau_{32}} - \left(\frac{1}{\tau_{21}} + \frac{1}{\tau_{nr}}\right)N_2 - \frac{(N_2 - N_1)}{\tau_{21}} n_{ph} \tag{5.95}$$

$$\frac{\partial N_1}{\partial t} = \left(\frac{1}{\tau_{21}} + \frac{1}{\tau_{nr}}\right)N_2 + \frac{(N_2 - N_1)}{\tau_{21}} n_{ph} - \frac{N_1}{\tau_{10}} \tag{5.96}$$

$$\frac{\partial N_0}{\partial t} = \frac{N_1}{\tau_{10}} - \frac{N_0}{\tau_P} \tag{5.97}$$

$$N_{tot} = N_0 + N_1 + N_2 + N_3, \tag{5.98}$$

where $N_i = N_i(\vec{r}, t)$, $i = 0, 1, 2, 3$ are the population number densities of the corresponding electron level, N_{tot} is the total number of electrons participating in the lasing process, $\gamma_{ij} \equiv 1/\tau_{ij}$

5.6. MICROSCOPIC THEORY OF RANDOM LASERS

are the transition rates from level i to j, and γ_{nr} is the non-radiative decay rate of the laser level 2. $\gamma_P \equiv 1/\tau_P$ is the transition rate due to homogeneous, constant, external pumping. Further $n_{ph} \equiv N_{ph}/N_{tot}$ is the photon number density, normalized to N_{tot}. In the stationary limit (i.e. $\partial_t N_i = 0$), the above system of equations can be solved for the population inversion $n_2 = N_2/N_{tot}$ to yield (γ_{32} and γ_{10} assumed to be large compared to all other rates)

$$n_2 = \frac{\gamma_P}{\gamma_P + \gamma_{nr} + \gamma_{21}(n_{ph} + 1)}. \tag{5.99}$$

The energy density $\Phi_{\epsilon\epsilon}$ calculated above relates to the diffusing photon number by the relation

$$n_{ph} = \int_{-\frac{1}{2}d}^{+\frac{1}{2}d} dZ' \Phi_{\epsilon\epsilon}(Z, Z') n_2(Z'). \tag{5.100}$$

The energy density correlation $\Phi_{\epsilon\epsilon}$ obeys the equation

$$-\frac{\partial^2}{\partial Z^2}\Phi_{\epsilon\epsilon} = \frac{1}{D}\left[\frac{D}{-\chi_d^2}\right]\Phi_{\epsilon\epsilon} + \frac{1}{D}\frac{\omega^2 \Delta G_0(\vec{Q}, \Omega)}{c_p^2 g_\omega^{(1)}\left[1 + \Delta(\omega)\right]}. \tag{5.101}$$

5.6.1 Microscopic Determination of the Amplification Rate of the Intensity

In order to study how the photonic gain predicted by the laser rate equations is incorporated into the transport theory, we compare the results of our microscopic selfconsistent transport theory of this chapter five, with the diffusive model of random lasing which has been outlined in chapter four before. The previously phenomenologically assumed diffusion equation for the photons reads

$$\frac{\partial n_{ph}}{\partial t} = D\frac{\partial^2 n_{ph}}{\partial Z^2} + \gamma_{21}(n_{ph} + 1)n_2 \tag{5.102}$$

and involves a term proportional to γ_{21}, which describes the intensity increase due to stimulated and spontaneous emission. This term originates from the semi-classical laser rate equations Eq. (5.96). The term $\gamma_{21}(n_{ph} + 1)n_2$ in Eq. (5.102) is equal to the amplification rate of the intensity and therefore not related to a specific diffusion model.
Now we compare these considerations with results of the exact microscopic approach. The microscopic analogy to the photon density is the energy density $\Phi_{\epsilon\epsilon}$. In the static limit ($\partial_t \Phi_{\epsilon\epsilon} = 0 \Leftrightarrow \Omega = 0$, and $\partial_t n_{ph} = 0$) we have to match the following two equations describing the photonic growth in either of the two models

$$-\frac{\partial^2}{\partial Z^2}\Phi_{\epsilon\epsilon} = \frac{1}{D}\left[\frac{D}{-\chi_d^2}\right]\Phi_{\epsilon\epsilon} + \frac{1}{D}\left[\frac{D}{-\chi_d^2}\right]$$
$$+ \left(\frac{1}{D}\frac{\omega^2 \Delta G_0(\vec{Q}, \Omega)}{c_p^2 g_\omega^{(1)}\left[1 + \Delta(\omega)\right]} - \frac{1}{D}\left[\frac{D}{-\chi_d^2}\right]\right) \tag{5.103}$$

$$-\frac{\partial^2 n_{ph}}{\partial Z^2} = \frac{1}{D}\gamma_{21}n_2 n_{ph} + \frac{1}{D}\gamma_{21}n_2 \tag{5.104}$$

The first term on the right hand side of both equations represents stimulated emission, the second term, however, accounts for the spontaneous emission. The third term on the r.h.s. of Eq. (5.103), represents light amplification which is due to the propagating intensity within a gain medium (without contributions from spontaneous emission). This term has of course no counterpart in the phenomenological model of chapter four, represented here by Eq. (5.104).
Now we directly compare the terms for amplification in both equations, and therefore write

$$\frac{1}{D}\gamma_{21}n_2 n_{ph} + \frac{1}{D}\gamma_{21}n_2$$
$$\triangleq \tag{5.105}$$
$$\frac{1}{D}\left[\frac{D}{-\chi_d^2}\right]\Phi_{\epsilon\epsilon} + \frac{1}{D}\left[\frac{D}{-\chi_d^2}\right] + \left(\frac{1}{D}\frac{\omega^2 \Delta G_0(\vec{Q}, \Omega)}{c_p^2 g_\omega^{(1)}\left[1 + \Delta(\omega)\right]} - \frac{1}{D}\left[\frac{D}{-\chi_d^2}\right]\right).$$

CHAPTER 5. SELFCONSISTENT MICROSCOPIC THEORY OF RANDOM LASING

We neglect all incoherent contributions resulting from spontaneous emission etc. and require the equality of the coherent grow rates, representing the stimulated emssion process. The grow rates of both equations can or must be identified with each other, thus they form the link of the selfconsistent transport theory to the laser rate equations

$$\gamma_{21} n_2 = \left[\frac{D}{-\chi_d^2} \right]. \tag{5.106}$$

By selfconsistent calculation of the grow term, we are able to determine the gain $\text{Im}\epsilon_s$ of the system only due to microscopic considerations and therefore do not need to rely on phenomenological assumptions.

To derive the solution the system is specified in terms of light frequency, scatterer radius, filling fraction of the scatterer and the dielectric functions for both background and scatterer's media. Then the diffusion coefficient D is selfconsistently calculated. With this diffusion constant, we solve the phenomenological laser model to obtain the inversion n_2, with this inversion we solve Eq. (5.106), which yields the imaginary part of the dielectric function. Finally with this newly calculated epsilon the calculation starts all over again, and is repeated until a convergence is found.

5.6. MICROSCOPIC THEORY OF RANDOM LASERS

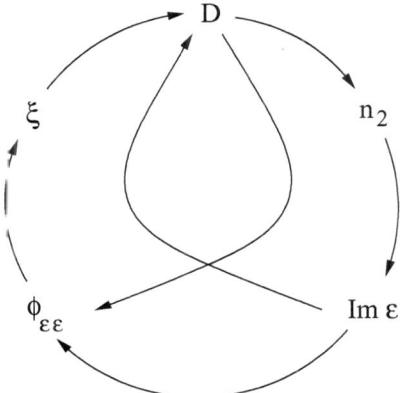

Figure 5.4: *Selfconsistent solving prozess of the microscopic transport theory coupled to the laser rate system. The general selfconstency is marked by the arrow crircle. The entangled arrows mark the seperate but coupled selfconsistencies of transport theory and laser rate equations.*

5.6.2 Numerics Procedure

Finally we want to summarize the procedure (see Fig. 5.4) which is used to gain the following results. The start is to solve the coupled laser rate equations selfconsistertly for appropriate start conditions. By gaining the inversion n_2 of the laser active niveau we derive the effective gain of the system, which is equal to the imaginary part of the dielectric constante ϵ and incorporate it in the selfconsistent solution of the above described transport theory including finite dimensions. From the structure of the energy density $\phi_{\epsilon\epsilon}$ we derive the correlation length which leads us to the selfconsistently calculated diffusion constant D. The diffusion processes and so the diffusion constant regulates the photon number per laser active atom which finally rules the inversion gain. So this process describes a selconstinecy process which includes two intrinsic selfconstint calculations sybolized by the entangled arrows.

5.7 Numerical Results and Discussion

In this section we finally present the numerical evaluations of our microscopic theory of random lasing. Due to the vast numerical data and large number of relevant quantities, we structure this section as follows: In the first part, subsection 5.7.1, we present numerical data for a system characterized by parameters, which are typical for experiments with ZnO powder. All quantities will be discussed in detail and also shown as function of the positional coordinate Z across the film. In subsection 5.7.2, we discuss the behavior of the correlation length, as the various system parameters are varied. And eventually, in subsection 5.7.3, we discuss the behavior of the diffusion constant as the system parameters, such as film width, filling fraction, light frequency and scattering strength, are changed. This last subsection will also reveal the influence of interference contributions on a random laser system.

5.7.1 Film of ZnO at 50 % Filling

In this first subsection we discuss a random laser, described by parameters which are typical for experiments. In particular, we chose $\text{Re}\,\epsilon_s = 10.4$, i.e. ZnO, a filling fraction $\nu = 50\%$, a frequency of $\omega/\omega_0 = 1.1$ where $\omega_0 = 2\pi c/r_0$ with the vacuum speed of light c and the scatterer radius r_0. The sample width is set to be $d = 40 r_0$. The coordinate Z is measured in units of the sample width, and pump rate is measured in units of the laser transition rate γ_{21}.
At first, the obtained photon density as a function of Z across the sample is shown in Fig. 5.5. For increasing pump rates the photon number increases strongly. In Fig. 5.6 the photon density at the sample surface, i.e. at $Z = 0.5d$ is displayed for increasing pump rates, for very large pump rates and also large photon numbers, the graph changes to a linear behavior, as also reported in experiments [28, 29].
The inversion density n_2 is displayed in Fig. 5.7 as a function of Z across the sample. Whereas the self-consistently obtained optical gain $\text{Im}\,\epsilon_s$ is shown in Fig. 5.8. A direct comparison of the atomic inversion n_2 and the self-consistently evaluated gain $\text{Im}\,\epsilon_s$ as a function of Z across the sample can be found in Fig. 5.9 for two typical, different pump rates. Additionally to difference in the magnitude, the gain $-\text{Im}\,\epsilon_s$ compared to the inversion displays a smaller slope towards the surface.

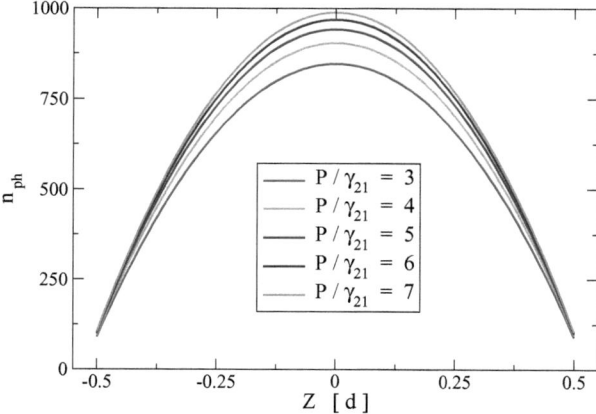

Figure 5.5: *The photon density n_{ph} is displayed as a function of Z across the sample width for pump rates $P/\gamma_{21} = 3, 4, 5, 6, 7$ in increasing order. The photon density represents the the number of photons per laser active atom. An increase of the photon number with increasing pumping is clearly observed.*

5.7. NUMERICAL RESULTS AND DISCUSSION

In Fig. 5.10, the correlation length ξ of the random laser is displayed, also for different pump rates. As explained above, the correlation length represents the average lateral extension within the (x, y)-plane for different values of Z. The volume it represents is the average mode extension within the film of ZnO. This is also illustrated in Fig. 5.11, where schematically the correlation volume of a lasing mode in the finite random laser sample is contrasted with the previously in chapter 3 derived correlation volume for infinite systems. The correlation length is measured in units of the sample width d. A general decrease of the correlation length with increasing pumping is observed across the sample. Furthermore, the correlation length assumes larger values, where the photon density is higher, i.e. towards the center of the film.

In Fig. 5.12 we present the correlation length ξ as a function of the pump rate, measured in units of the transition rate γ_{21}. The correlation length calculated by coupling the laser rate equation to the microscopic transport theory of light in presence of gain, exhibits a decreasing behavior with an increasing pump rate. To study the behavior of this decrease we plot the correlation length versus the inverse square root of the pump rate, just as we had done it for phenomenological model in chapter 4. The result is shown in Fig. 5.13. The blue line in Fig. 5.13 serves as a guide to the eye, emphasizing the linear behavior of the correlation length in this plot. This clearly reveals an inverse square root behavior of ξ as a function of pumping above threshold.

After the discussion of the evaluated quantities directly concerning the lasing intensity, let us now turn to the transport characteristics in such systems. In Fig. 5.14 we display the full, self-consistently calculated diffusion coefficient D containing all interference effects as a function of Z across the sample for various pump rates. The unit of D is $r_0 c$, where r_0 is the radius of the scatterers and c is again the vacuum speed of light. With increasing pumping an increase of the diffusion coefficient is observed. Close to the surfaces, diffusion is stronger. The reason for this is, that there, the possibility for multiple scattering and also for forming closed loops is suppressed. Both processes would tend to slow down the light propagation. Since they are missing however, light transport speeds up. Additionally, close to the surfaces the optical gain is also stronger, because the photon number is smaller there. This also tends to increase the diffusion constant.

In Fig. 5.15, we also show for completeness the bare diffusion coefficient D_0 as a function of Z for the same pump rates as in Fig. 5.14. The bare diffusion coefficient measures how fast a given mode propagates within the sample and out of it given that no interference effects take place. The unit of D_0 is also $r_0 c$. A similar behavior of D_0 as compared to D is found.

A detailed comparison of bare diffusion coefficient D_0 (dashed line) and the full diffusion coefficient D (solid line) as a function of Z across the sample width for a typical pump rate $P = 3\gamma_{21}$ can be found in Fig. 5.16.

Finally, the ratio D/D_0 is displayed in Fig. 5.17. The ratio D/D_0 measures the deviation of the full diffusion constant including interference effects from a purely diffusive system with no Cooperon contributions. Therefore providing a measure of the influence of interference effects (Cooperons) on the transport in the system. With increasing pumping the ratio decreases, indicating a growth of interference effects occurring in the system as the optical gain increases. The strong decreases towards the surfaces is driven by the increase of Im ϵ_s, as shown in Fig. 5.8. Roughly speaking, the self-interference effects diminish the diffusion constant by 5% as compared to a purely diffusive system.

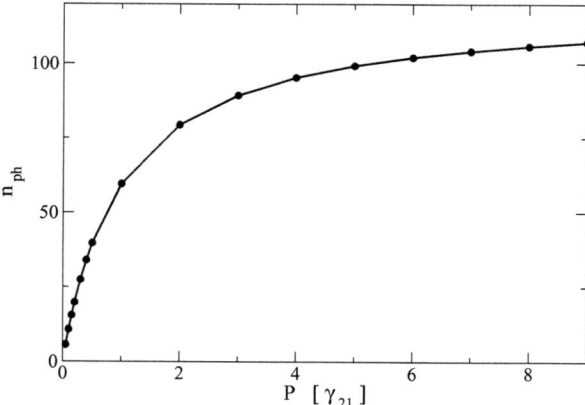

Figure 5.6: *Photon density n_{ph} as a function of the pump rate, measured in units of the transition rate γ_{21}, at the sample surface, i.e. $Z = 0.5d$. For large pump rates (and large photon numbers) the photon number exhibits a linear behavior, as also reported in experiments [28, 29].*

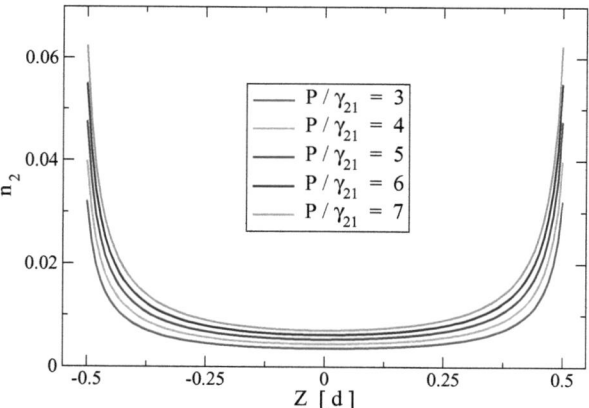

Figure 5.7: *The inversion density n_2 is shown as a function of Z across the sample width for pump rates $P/\gamma_{21} = 3, 4, 5, 6, 7$ in increasing order. The inversion density represents the number of inverted atoms divided by the total number of invertible atoms, the width is measured in units of the sample width, and pump rate is measured in units of the transition rate γ_{21}. An increase of the inversion density with increasing pumping indicates, that the gain is not saturated yet.*

5.7. NUMERICAL RESULTS AND DISCUSSION

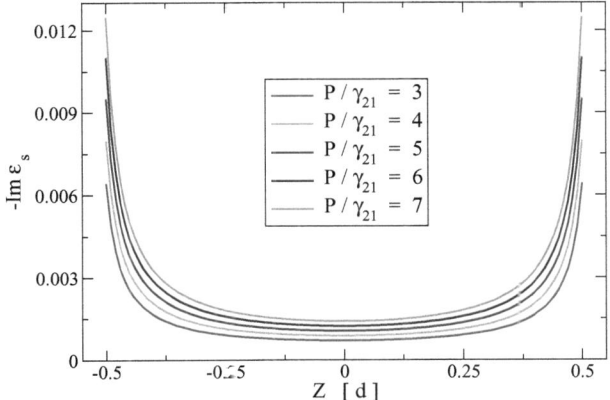

Figure 5.8: *We show the self-consistently obtained optical gain* $\mathrm{Im}\,\epsilon_s$ *as a function of Z across the sample width for pump rates* $P/\gamma_{21} = 3, 4, 5, 6, 7$ *in increasing order. The optical gain represented by the imaginary part of the dielectric function of scatterers* ϵ_s *translates the electronic or atomic inversion density into an amplification of light intensity in the disordered sample. An increase of the optical gain with increasing pumping is observed, analog to the behavior of atomic inversion, shown in Fig. 5.7.*

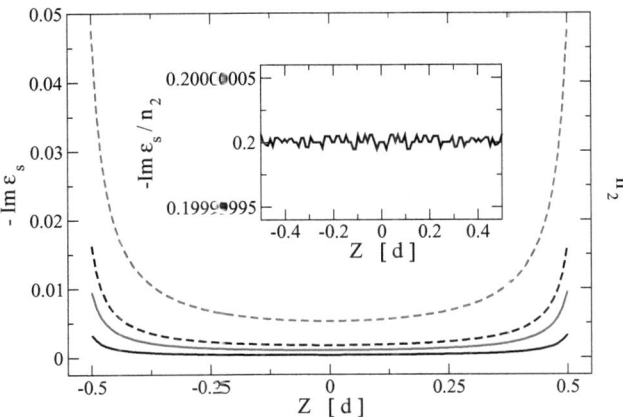

Figure 5.9: *The direct comparison is drawn between the self-consistently obtained optical gain* $\mathrm{Im}\,\epsilon_s$ *(solid lines) with the corresponding calculated atomic inversion* n_2 *(dashed lines) as a function of Z across the sample width. Black curves are evaluated for* $P = 1\gamma_{21}$ *and the red ones for* $P = 5\gamma_{21}$. *In the inset we see the relation between both quantities which is almost perfectly constant.*

CHAPTER 5. SELFCONSISTENT MICROSCOPIC THEORY OF RANDOM LASING

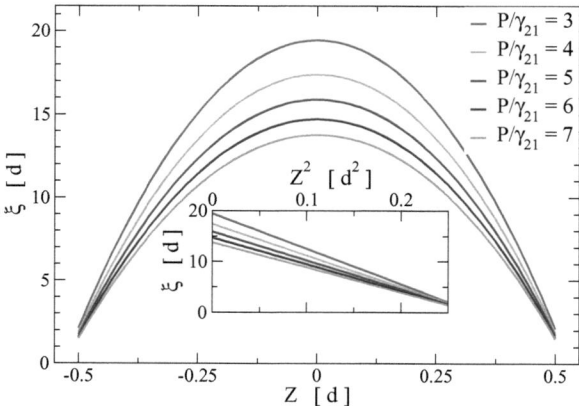

Figure 5.10: *The correlation length ξ is shown as a function of Z across the sample width for pump rates $P/\gamma_{21} = 3, 4, 5, 6, 7$ in increasing order. The correlation length represents the average lateral extension within the (x, y)-plane for different values of Z. The volume it represents is the average mode extension within the film of ZnO. Also compare with Fig. 5.11 below. The correlation length is measured in units of the sample width d. A decrease of the correlation length with increasing pumping is observed across the sample. Furthermore, the correlation length assumes larger values, where the photon density is higher, i.e. towards the center of the film. The inset shows a better resolution plot near the surface. We find that the correlation length shows an square dependence to the depth in the sample as can be seen in the inset.*

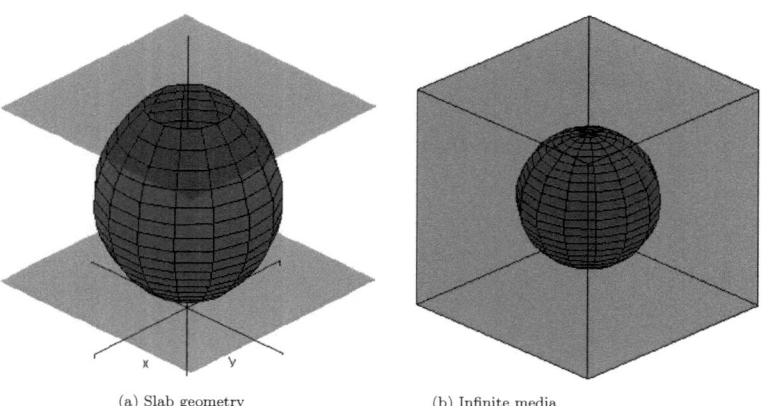

(a) Slab geometry (b) Infinite media

Figure 5.11: *Illustration of the correlation volume, described by ξ.*
(a) Correlation volume of a diffusing/lasing mode within a random laser film. This cigar-shaped volume connects the two surfaces of the sample and is described via the correlation length $\xi(Z)$ within the (x, y)-plane. Numerical evaluation are displayed in Fig. 5.10 above.
(b) In the case of an infinite medium the correlation volume of a lasing mode is a sphere with radius R_{max} as derived and discussed in chapter 3 and illustrated here.

5.7. NUMERICAL RESULTS AND DISCUSSION

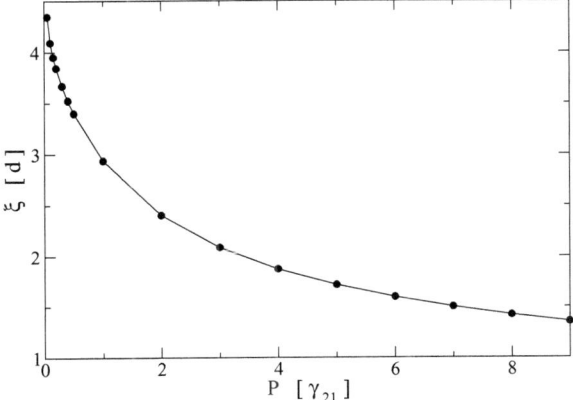

Figure 5.12: *The correlation length ξ is displayed as a function of the pump rate, measured in units of the transition rate γ_{21}, at the sample surface, i.e. $Z = 0.5d$. The sample width is set to be $d = 40r_0$, where r_0 is the radius of the scatterers, the frequency is $\omega/\omega_0 = 1.1$ and the filling fraction is $\nu = 50\%$.*

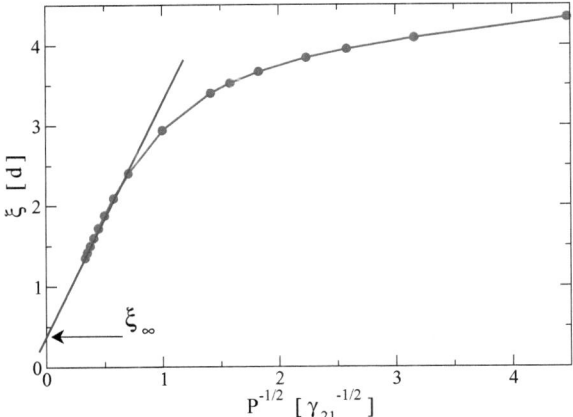

Figure 5.13: *Correlation length ξ shown as a function of the inverse square root of the pump rate, measured in units of the transition rate γ_{21}, at the sample surface, i.e. $Z = 0.5d$, c.f. Fig. 5.11. The sample width is set to be $d = 40r_0$, where r_0 is the radius of the scatterers, the frequency is $\omega/\omega_0 = 1.1$ and the filling fraction is $\nu = 50\%$.*
The blue line serves as a guide to the eye, emphasizing the linear behavior of the correlation length in this plot. This clearly reveals an inverse square root behavior of ξ as a function of pumping above threshold. We find a finite correlation length for unlimited pump strengths.

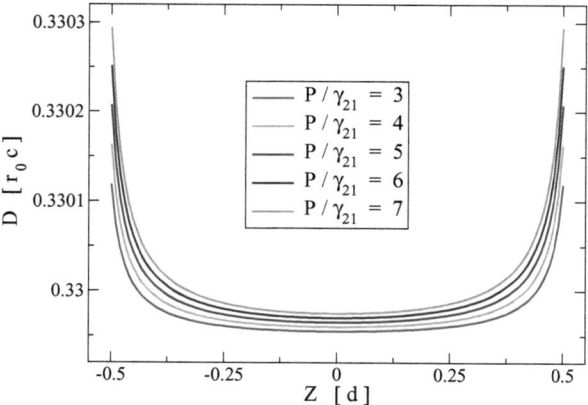

Figure 5.14: *The full, self-consistently calculated diffusion coefficient D is plotted as a function of Z across the sample width for pump rates $P/\gamma_{21} = 3, 4, 5, 6, 7$ in increasing order. The diffusion coefficient measures how fast a given mode propagates within the sample and out of it. The unit of D is $r_0 c$, where r_0 is the radius of the scatterers and c is the vacuum speed of light. The coordinate Z is measured in units of the sample width d, and the pump rate is measured in units of the transition rate γ_{21}. With increasing pumping an increase of the diffusion coefficient is observed. Close to the surfaces, diffusion is stronger, because there the possibility for multiple scattering and also for forming closed loops is suppressed. Both processes would tend to slow down the light propagation.*

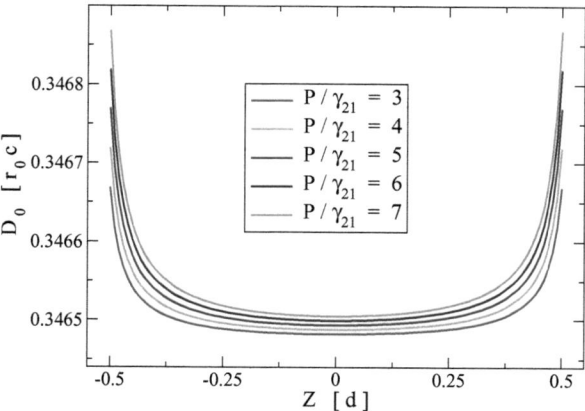

Figure 5.15: *Bare diffusion coefficient D_0 as a function of Z across the sample width for pump rates $P/\gamma_{21} = 3, 4, 5, 6, 7$ in increasing order. The bare diffusion coefficient measures how fast a given mode propagates within the sample and out of it given that NO interference effects take place. The unit of D_0 is $r_0 c$. The coordinate Z is measured in units of the sample width d, and pump rate is measured in units of the transition rate γ_{21}. With increasing pumping an increase of the bare diffusion coefficient is observed. Close to the surfaces, bare diffusion is stronger, because there the possibility for multiple scattering is suppressed.*

5.7. NUMERICAL RESULTS AND DISCUSSION

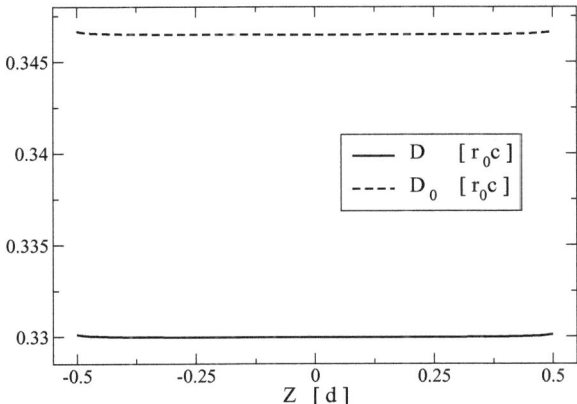

Figure 5.16: *We compare the bare diffusion coefficient D_0 (dashed line) to the full diffusion coefficient D (solid line) as a function of Z across the sample width for pump rates $P/\gamma_{21} = 3\gamma_{21}$.*

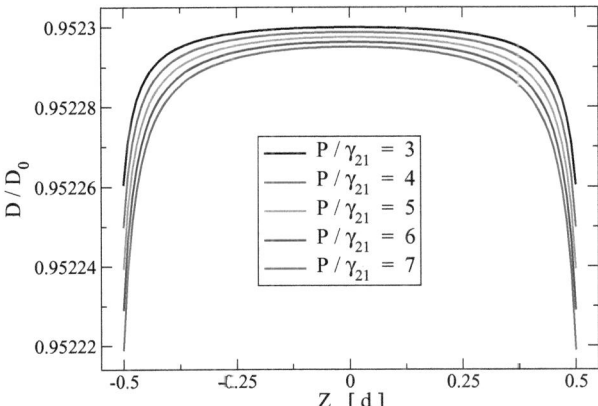

Figure 5.17: *The full and self-consistent diffusion constant D in units of the bare diffusion constant D_0 is displayed as a function of Z across the sample width for pump rates $P/\gamma_{21} = 3, 4, 5, 6, 7$ in increasing order. The ratio D/D_0 measures the deviation of the full diffusion constant including interference effects from a purely diffusive system with no Cooperons contributions. The sample width is set to be $d = 40r_0$ where r_0 is the radius of the scatterers, the frequency is $\omega/\omega_0 = 1.1$ and the filling fraction is $\nu = 50\%$.*

With increasing pumping the ratio decreases, indicating a growth of interference effects occurring in the system as the optical gain increases. The strong decreases towards the surfaces is driven by the increase of $\text{Im}\,\epsilon_s$, as shown in Fig. 5.8.

Roughly speaking, the self-interference effects diminish the diffusion constant by 5% as compared to a purely diffusive system.

5.7.2 Behavior of the Correlation Length for Various System Parameters

In this subsection we discuss the behavior of the correlation length ξ as the various system parameters are changed. The reference system to which we always compare is the one, presented in detail in the previous subsection and characterized by the experimentally relevant parameters: $\text{Re}\,\epsilon_s = 10.4$, i.e. ZnO, filling fraction $\nu = 50\%$, a frequency of $\omega/\omega_0 = 1.1$ where $\omega_0 = 2\pi c/r_0$ with the vacuum speed of light c and the scatterer radius r_0. The sample width is $d = 40r_0$. The coordinate Z is measured in units of the sample width, and the pump rate is measured in units of the laser transition rate γ_{21}.

The first parameter to be changed is the width of the lasing film, which is relatively easily accomplished in experiments. In Fig. 5.18 and also in Fig. 5.19 we compare two systems one of width $d = 40r_0$, as explained above, and the other one at a larger size of $d = 80r_0$. In Fig. 5.18, we show the correlation length ξ as a function of the pump rate, measured in units of the transition rate γ_{21}, at the sample surface, i.e. $Z = 0.5d$. Whereas in Fig. 5.19 the same ξ is now plotted as a function of the inverse square root of the pump rate. In this graph one again clearly observes the inverse square root decrease of the correlation length as a function of increasing pump rate. This is in this particular plot indicated by the linear behavior for small $P^{-1/2}$. The blue straight lines serve as a guide to the eye, emphasizing the linear behavior of the correlation length in this plot, as discussed. In general, one observes an increase of the correlation length at the surface with increasing film widths, which is related to the fact, the correlation length tends to increase with increasing photon number, as shown in Fig. 5.10. The larger the width, the larger is the photon density, because then the surface to volume ratio decreases and hence the loss through the surfaces also decreases with increasing width.

The next parameter to change is the filling fraction of the scatterers immersed in the host medium. The corresponding comparisons are found in Fig. 5.20 and Fig. 5.21. The system is characterized by the standard settings given above, except for filling fractions of $\nu = 35\%, 40\%, 45\%, 50\%, 60\%$. In Fig. 5.20 the correlation length as function of pump rate is shown, whereas in Fig. 5.21 the same ξ is now displayed as a function of the inverse square root of the pump rate. A decrease of the correlation length with increasing filling fraction is observed, while at the same time the inverse square root decay can be observed for all filling fraction as seen in Fig. 5.21. It is to be noted, that for increasing filling fraction close to and above values of $\nu \sim 50\%$ the change of ξ becomes more and more negligible. This behavior reflects the fact, that in this range the limit of densest-packing of spheres is reached.

In Fig. 5.22 and Fig. 5.23, the random laser system is operated with two different laser light frequencies, and the resulting correlation length is shown as function of pump rate Fig. 5.22, and as a function of the inverse square root of the pump rate, Fig. 5.23. A decrease of the frequency yields somewhat larger values of ξ. The difference of the two curves is largest for small pump rates, and also the onset of the linear behavior seems to occur at a different pumping. The difference of the two frequencies is their position relative to a Mie-resonance and will become clearer in the next subsection, discussing the behavior of the diffusion constant.

The last parameter which is changed in the random laser is the scattering strength of the individual scatteres embedded in the host medium and hence forming the random medium. In Fig. 5.24 we display the correlation length ξ as a function of the pump rate, measured in units of the transition rate γ_{21}, at the sample surface, i.e. $Z = 0.5d$, as a comparison for real parts of the dielectric constant given by $\epsilon_s = 10.4$ (ZnO-reference) and $\epsilon_s = 9.0$. And in Fig. 5.25, the same ξ is now displayed as a function of the inverse square root of the pump rate. The rather large effect on the correlation length can again be traced back to the behavior of the Mie-resonances, which change their position as function of scattering strength. This is shown in Fig. 5.30 and Fig. 5.31 in the next subsection, and will also be discussed there.

5.7. NUMERICAL RESULTS AND DISCUSSION

Different Widths

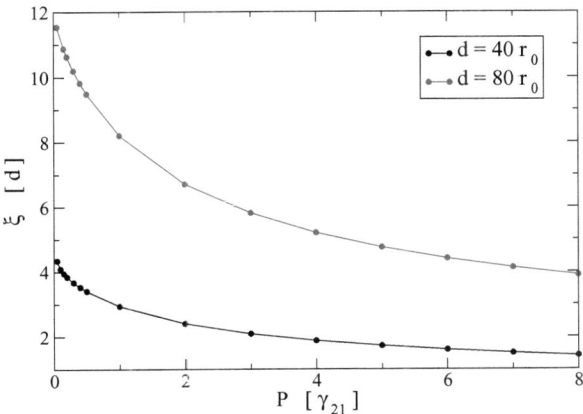

Figure 5.18: *Correlation length ξ as a function of the pump rate, measured in units of the inverse square root of the transition rate γ_{21}, at the sample surface, i.e. $Z = 0.5d$ The sample width is set to be $d = 40r_0$ (black line, also displayed in Fig. 5.12) and $d = 80r_0$ (red line), where r_0 is the radius of the scatterers, the frequency is $\omega/\omega_0 = 1.1$ and the filling fraction is in both samples $\nu = 50\%$.*

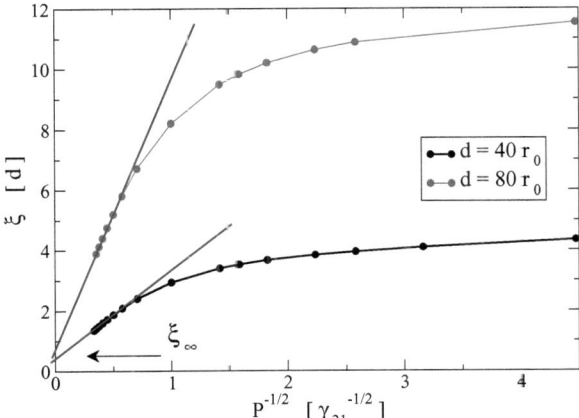

Figure 5.19: *Correlation length ξ as a function of the inverse square root of the pump rate, measured in units of the inverse square root of the transition rate γ_{21}, at the sample surface, i.e. $Z = 0.5d$, c.f. Fig. 5.17.*
The blue lines serve as a guide to the eye, emphasizing the linear behavior of the correlation length in this plot. This clearly reveals an inverse square root behavior of ξ as a function of pumping above threshold.

Different Filling Fractions

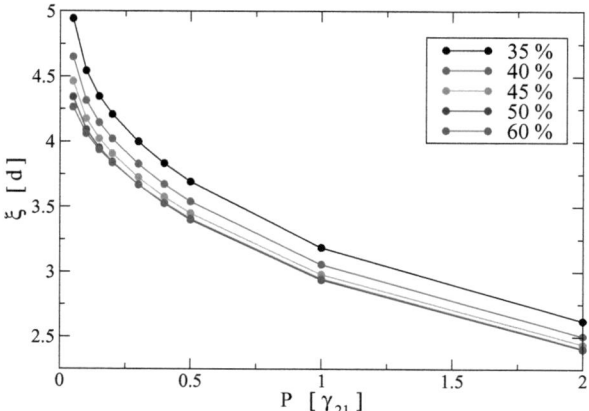

Figure 5.20: *The correlation length ξ is shown as a function of the pump rate, measured in units of the transition rate γ_{21}, at the sample surface, i.e. $Z = 0.5d$. The filling fraction is changed from $\nu = 35\%$ up to $\nu = 60\%$ as indicated in the legend. The reference data for $\nu = 50\%$ have also been presented above in Fig. 5.12.*

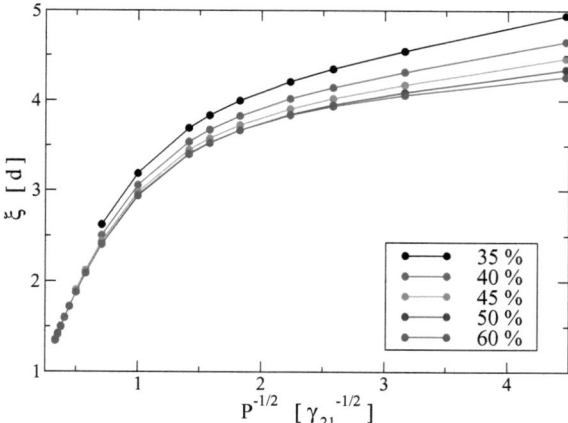

Figure 5.21: *The correlation length ξ is displayed as a function of the inverse square root of the pump rate, measured in units of the transition rate γ_{21}, at the sample surface, i.e. $Z = 0.5d$, c.f. Fig. 5.20. The filling fraction is changed from $\nu = 35\%$ up to $\nu = 60\%$ as indicated in the legend. The reference data for $\nu = 50\%$ have also been presented above in Fig. 5.12. The linear behavior found here for small $P^{-1/2}$ reveals the inverse square root decay of the correlation length as a function of increasing pump rate.*

5.7. NUMERICAL RESULTS AND DISCUSSION

Different Light Frequencies ω

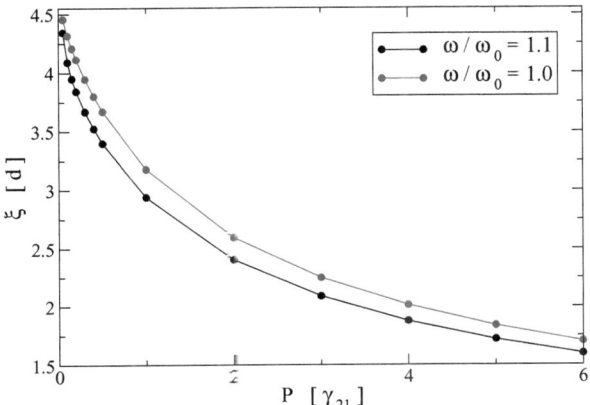

Figure 5.22: *Correlation length ξ as a function of the pump rate, measured in units of the transition rate γ_{21}, at the sample surface, i.e. $Z = 0.5d$. The sample width is set to be $d = 40r_0$, where r_0 is the radius of the scatterers, the frequency is chosen as $\omega/\omega_0 = 1.1$ and 1.0 respectively as indicated in the legend. For both systems the filling fraction is $\nu = 50\%$. The reference data for $\omega/\omega_0 = 1.1$ have also been presented above in Fig. 5.12.*

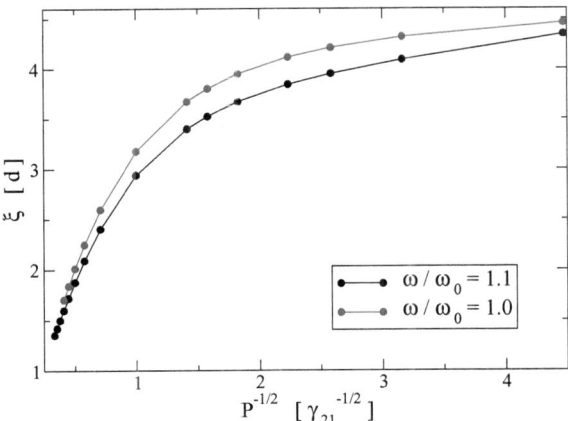

Figure 5.23: *Correlation length ξ as a function of the inverse square root of the pump rate, measured in units of the inverse square root of the transition rate γ_{21}, at the sample surface, i.e. $Z = 0.5d$, c.f. Fig. 5.22. The sample width is set to be $d = 40r_0$, where r_0 is the radius of the scatterers, the frequency is $\omega/\omega_0 = 1.1$ and 1.0 respectively as indicated in the legend. For both systems the filling fraction is $\nu = 50\%$.*

Different scattering strength Re ϵ_s

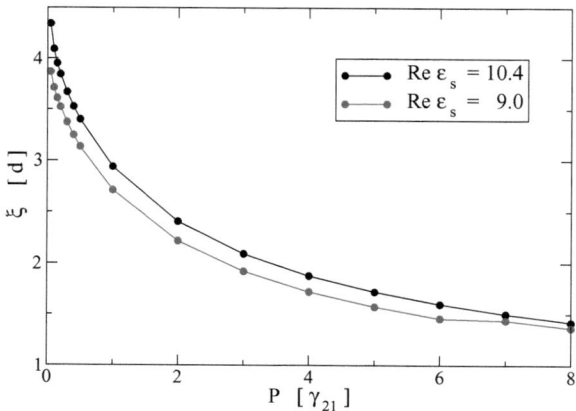

Figure 5.24: *Correlation length ξ as a function of the pump rate, measured in units of the transition rate γ_{21}, at the sample surface, i.e. $Z = 0.5d$. The sample width is set to be $d = 40r_0$, where r_0 is the radius of the scatterers, the frequency is chosen as $\omega/\omega_0 = 1.1$. The real part of the dielectric function of the scatterers is chosen as $\epsilon_s = 10.4$ (ZnO) and as $\epsilon_s = 9.0$ respectively as indicated in the legend. The filling fraction is $\nu = 50\%$ for both systems. The reference data for $\epsilon_s = 10.4$ have also been presented above in Fig. 5.12.*

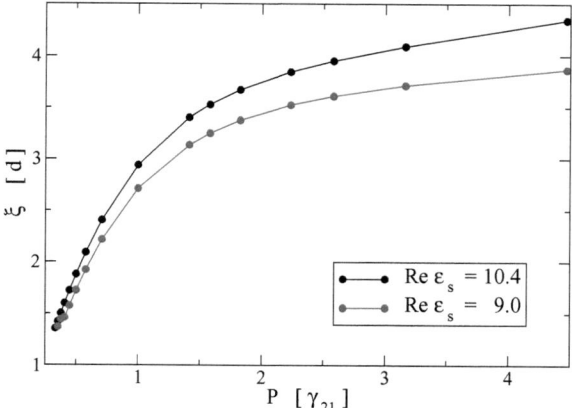

Figure 5.25: *Correlation length ξ as a function of the inverse square root of the pump rate, measured in units of the transition rate γ_{21}, at the sample surface, i.e. $Z = 0.5d$, c.f. Fig. 5.24. The sample width is set to be $d = 40r_0$, where r_0 is the radius of the scatterers, the frequency is $\omega/\omega_0 = 1.1$. The real part of the dielectric function of the scatterers is chosen as $\epsilon_s = 10.4$ (ZnO) and as $\epsilon_s = 9.0$ respectively as indicated in the legend. The filling fraction is $\nu = 50\%$ for both systems.*

5.7.3 Behavior of the Diffusion Coefficient for Various System Parameters

In this subsection the discuss the behavior of the self-consistently calculated diffusion constant D for a random lasing system of finite width as the various system parameters are changed. The reference system to which we always compare is the one, presented in detail in the first subsection and characterized by the experimentally relevant parameters: $\mathrm{Re}\,\epsilon_s = 10.4$, i. e. ZnO, filling fraction $\nu = 50\%$, a frequency of $\omega/\omega_0 = 1.1$ where $\omega_0 = 2\pi c/r_0$ with the vacuum speed of light c and the scatterer radius r_0. The sample width is $d = 40r_0$. The coordinate Z is measured in units of the sample width, and pump rate is measured in units of the laser transition rate γ_{21}.
The first parameter to be changed is again the width of the lasing film, which is relatively easily accomplished in experiments. In Fig. 5.26 and also in Fig. 5.27 we compare two systems one of width $d = 40r_0$, as explained above, and the other one at a larger size of $d = 80r_0$.
In Fig. 5.26, we show the full, self-consistently calculated diffusion coefficient D as a function of Z across the sample width for a typical pump rate of $P/\gamma_{21} = 2$. Whereas in Fig. 5.27 the full and self-consistent diffusion constant D in units of the bare diffusion constant D_0 is displayed, i.e. the ratio D/D_0, as a function of the coordinate Z across the film width, Z is measured in units of the film width d. The diffusion coefficient measures how fast a given mode propagates within the sample and out of it, and can experimentally measured. Here the unit of D is $r_0 c$, where r_0 is the radius of the scatterers and c is the vacuum speed of light. The ratio D/D_0 measures the deviation of the full diffusion constant including interference effects from a purely diffusive system with no Cooperons contributions. Fig. 5.26 shows that with increasing sample width, the diffusion coefficient remains practically unchanged, since the relative change observed in the graph is of the same order as the accuracy involved in the numerical evaluation, i.e. 10^{-5}. Also the ratio D/D_0, Fig. 5.27, remains practically unchanged when the size of the lasing film is doubled. This had to be expected since the diffusion is mainly governed by bulk properties, which of course are invariant under a change of system size. This however is in contrast to for instance the correlation length, which changes significantly as the system size changes.
The next parameter to change is the filling fraction of the scatterers immersed in the host-medium. The corresponding comparisons are found in Fig. 5.28 and Fig. 5.29. The system is characterized by the standard settings given above, except for filling fractions of $\nu = 40\%$ and $\nu = 50\%$. In Fig. 5.28 the full, self-consistently calculated diffusion coefficient D as a function of Z across the sample width for pump rate $P/\gamma_{21} = 2$ is shown. With increasing sample width, the diffusion coefficient is decreasing, which is due to the increasing amount of multiple scattering in the more densely packed system. In Fig. 5.29 the ratio D/D_0 is displayed for the same system. As discussed above, also the ratio D/D_0 decreases with increasing filling fraction ν. Since the stronger scattering also favors interference effects. Therefore the localization effects, i.e. Cooperon contributions, are enhanced. This is again in contrast to the behavior of the correlation length in Fig. 5.20, which is only weakly affected by a change in the filling fraction.
The next parameter we wish to study is the light frequency ω. To this end, we first present in Fig. 5.30 the full diffusion coefficient D as a function of light frequency for the two scattering strengths $\epsilon_s = 10.4$ and $\epsilon_s = 9.0$ as needed below and also presented in the previous subsection. Furthermore, in Fig. 5.31 we show the diffusion coefficient D in units of the bare diffusion constant D_0 which reveals the amplifying character of the Mie-resonances on the Cooperon contributions, i.e. localization effects are enhanced. Additionally, the two frequencies $\omega/\omega_0 = 1.0$ and $\omega/\omega_0 = 1.1$ are indicated at which the random laser will be operated. In both graphs, the Mie-resonances and their decreasing influence on the diffusion constant is clearly visible. These graphs serve as reference for the lasing systems below.
Fig. 5.32 shows the diffusion coefficient D of the random laser comparing frequencies $\omega/\omega_0 = 1.1$ and $\omega/\omega_0 = 1.0$. With a lower frequency, i.e. a frequency which is closer to an internal Mie-resonance, the light propagation is skewed down since light now remains partially trapped in such resonances as illustrated in Fig. 5.31. Fig. 5.33 demonstrates, that the interference effects are the stronger increased the closer the light frequency is to the resonance. The effects originating from interference phenomena are more than a factor of 4 stronger for $\omega/\omega_0 = 1.0$.
The last parameter which is changed in the random laser is the scattering strength of the individual

scatteres embedded in the host medium and hence forming the random medium. In Fig. 5.34 and Fig. 5.35 we display the diffusion coefficient D (D/D_0) as a function of Z across the sample width for pump rate $P/\gamma_{21} = 2$ as a comparison for real parts of the dielectric constant given by $\text{Re}\,\epsilon_s = 10.4$ (ZnO-reference) and $\text{Re}\,\epsilon_s = 9.0$. The rather large influence of D and D/D_0 when changing $\text{Re}\,\epsilon_s$, is solely due to the fact that the Mie-resonances shift as the scattering is changed, and therefore by comparing the same frequency for both systems results in one off-resonant measurement and in one resonant measurement. That explains the observed behavior. With decreasing ϵ_s, in Fig. 5.34, the position of the internal Mie-resonances is changed along the frequence axis. This effect is illustrated above in Fig. 5.31. Therefore the light propagation is slowed down since light now remains partially trapped in such resonances.

5.7. NUMERICAL RESULTS AND DISCUSSION

Different Widths

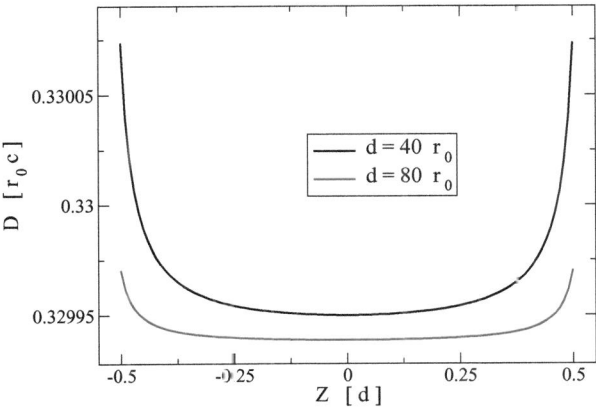

Figure 5.26: *The full, self-consistently calculated diffusion coefficient D as a function of Z across the sample width for pump rate $P/\gamma_{21} = 2$. The sample width is set to be $d = 40r_0$ and $d = 80r_0$ respectively as indicated in the legend, where r_0 is the radius of the scatterers. The reference data for $\epsilon_s = 10.4$ (ZnO) have also been presented above in Fig. 5.14.*

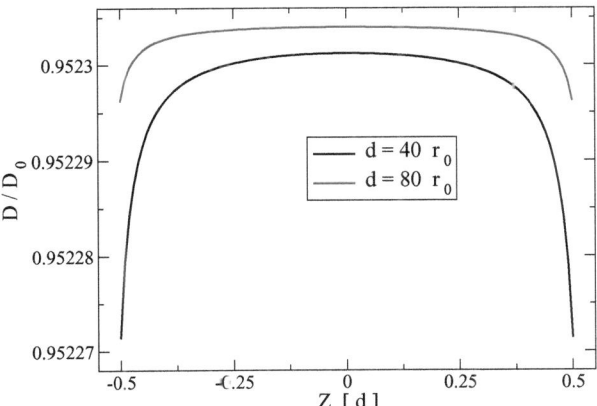

Figure 5.27: *The full and self-consistent diffusion constant D in units of the bare diffusion constant D_0 is shown as a function of Z across the sample width for pump rate $P/\gamma_{21} = 2$. The sample width is set to be $d = 40r_0$ and $d = 80r_0$ respectively as indicated in the legend, where r_0 is the radius of the scatterers. As discussed in the text, also the ratio D/D_0 remains practically unchanged, since the relative change observed in the graph is of the same order as the accuracy involved in the numerical evaluation. This implies that interference effects do not change as a function of the sample width.*

5.7.4 Different Filling Fractions

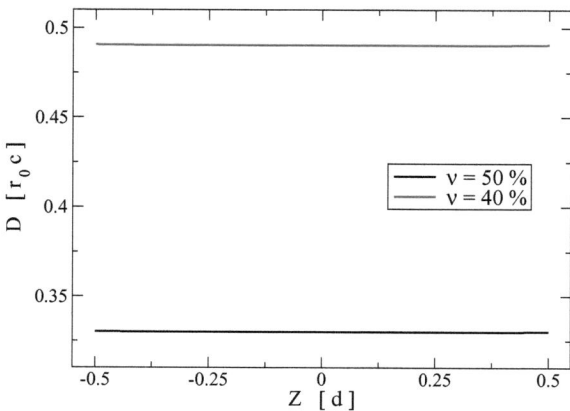

Figure 5.28: *The full, self-consistently calculated diffusion coefficient D is displayed as a function of Z across the sample width for pump rate $P/\gamma_{21} = 2$. With increasing sample width, the diffusion coefficient is decreasing, which is due to the increasing amount of multiple scattering in the more densely packed system. The reference data for $\epsilon_s = 10.4$ (ZnO) have also been presented above in Fig. 5.14.*

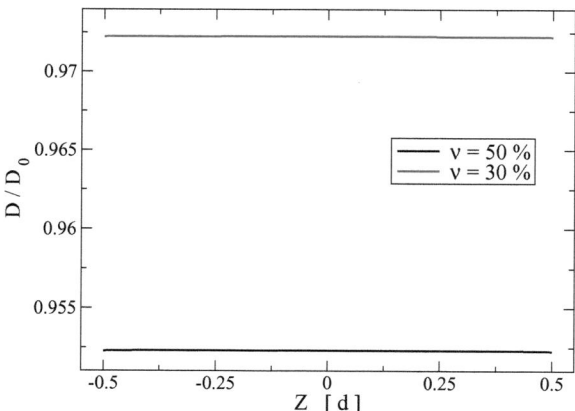

Figure 5.29: *Full and self-consistent diffusion constant D in units of the bare diffusion constant D_0 as a function of Z across the sample width for pump rate $P/\gamma_{21} = 2$. The filling fraction is $\nu = 50\%$ and $\nu = 40\%$ respectively as indicated in the legend. As discussed in the text, also the ratio D/D_0 decreases with increasing filling fraction ν, since a stronger scattering also favors interference effects.*

5.7. NUMERICAL RESULTS AND DISCUSSION

Different Light Frequencies ω

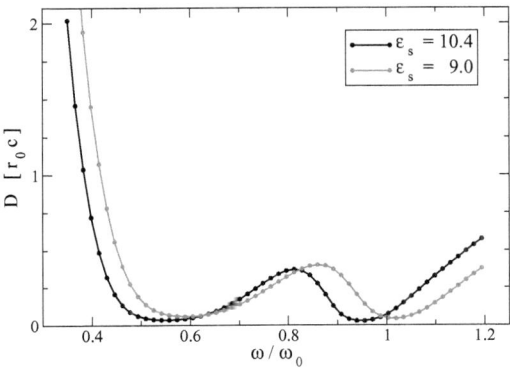

Figure 5.30: *Full, self-consistently calculated diffusion coefficient D in units of the bare diffusion constant D_0 for pump rate $P/\gamma_{21} = 0$ as a function of light frequency. The Mie-resonances and their decreasing influence on the diffusion constant is clearly visible. This graph serves as reference for the lasing systems below.*

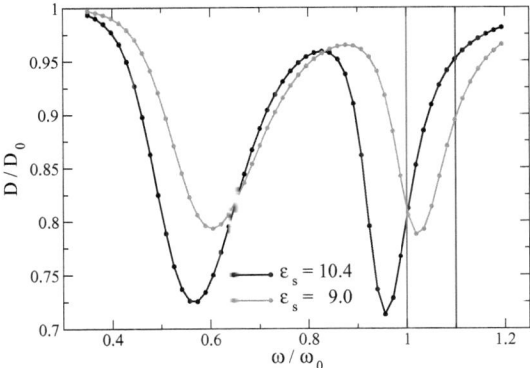

Figure 5.31: *Full, self-consistently calculated diffusion coefficient D for pump rate $P/\gamma_{21} = 0$ as a function of light frequency. The red and blue line represent the frequencies $\omega/\omega_0 = 1.0$, $\omega/\omega_0 = 1.1$ respectively, at which the lasing system has been evaluated in the following. The frequency $\omega/\omega_0 = 1.0$ is located within the Mie-resonance, whereas the $\omega/\omega_0 = 1.1$ is off-resonance. This black graph serves as reference for the lasing systems below. The displayed green line for $\epsilon_s = 9.0$ will be discussed below, please note here that at frequency $\omega/\omega_0 = 1.1$ the interference contributions are stronger as compared to the ZnO system with $\epsilon_s = 10.4$, an effect which is purely due to the different position of the Mie-resonances along the frequence axis.*

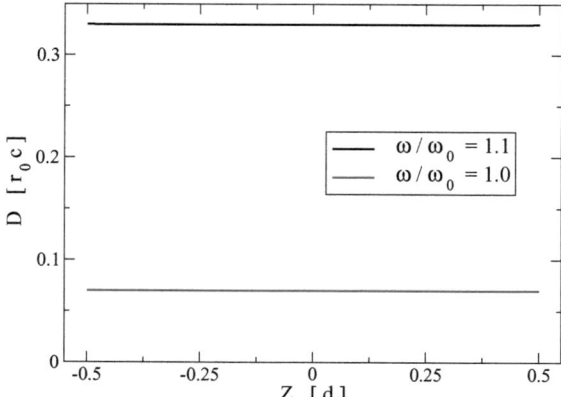

Figure 5.32: The full, self-consistently calculated diffusion coefficient D is displayed as a function of Z across the sample width for pump rate $P/\gamma_{21} = 2$. The light frequency is $\omega/\omega_0 = 1.1$ and $\omega/\omega_0 = 1.0$ respectively, as indicated in the legend. At the lower frequency, i.e. a frequency which is closer to an internal Mie-resonance, the light propagation is slowed down since light now remains partially trapped in such resonances. This effect is illustrated above in Fig. 5.31. The reference data for $\omega/\omega_0 = 1.1$ have also been presented above in Fig. 5.14.

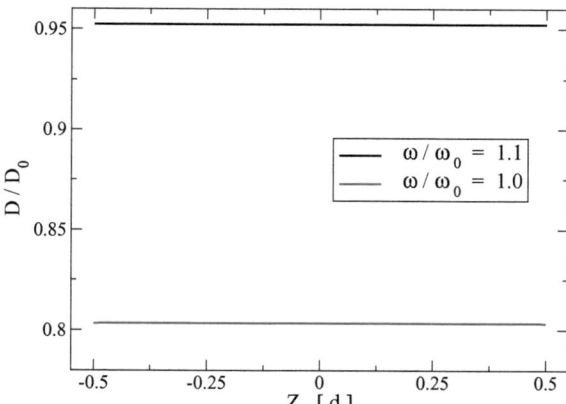

Figure 5.33: The full and self-consistent diffusion constant D in units of the bare diffusion constant D_0 is shown as a function of Z across the sample width for pump rate $P/\gamma_{21} = 2$. The light frequency is $\omega/\omega_0 = 1.1$ and $\omega/\omega_0 = 1.0$ respectively, as indicated in the legend. At the lower frequency, i.e. a frequency which is closer to an internal Mie-resonance, c.f. Fig. 5.31, an increase of interference effects is observed from roughly 4.8 % up to ca. 20.0 %. The effects originating from interference phenomena are therefore more than a factor of 4 stronger.

5.7. NUMERICAL RESULTS AND DISCUSSION

Different scattering strength Re ϵ_s

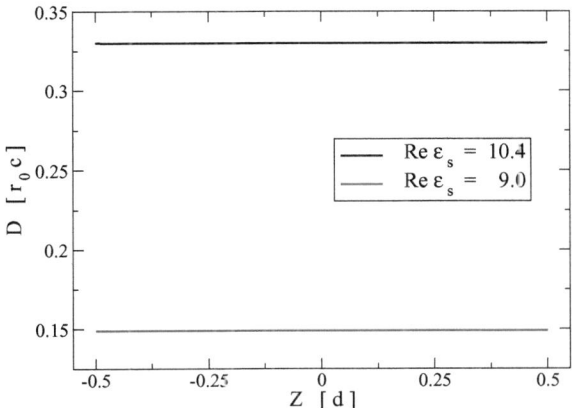

Figure 5.34: *Full, self-consistently calculated diffusion coefficient D as a function of Z across the sample width for pump rate $P/\gamma_{21} = 2$. The real part of the dielectric function of the scatterers is chosen as $\epsilon_s = 10.4$ (ZnO) and as $\epsilon_s = 9.0$ respectively as indicated in the legend.*

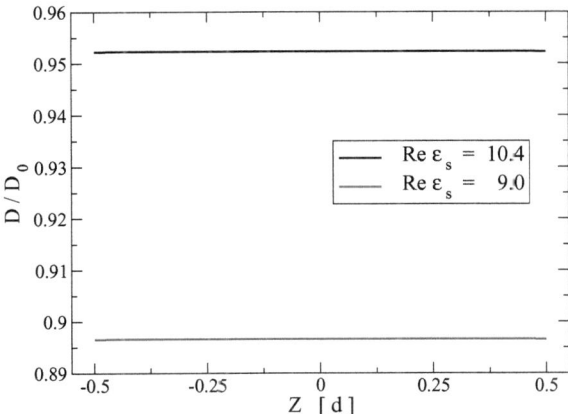

Figure 5.35: *The full and self-consistent diffusion constant D in units of the bare diffusion constant D_0 is shown as a function of Z across the sample width for pump rate $P/\gamma_{21} = 2$. The real part of the dielectric function of the scatterers is chosen as $\epsilon_s = 10.4$ (ZnO) and as $\epsilon_s = 9.0$ respectively as indicated in the legend. As discussed in the caption of Fig. 5.24, also c.f. Fig. 5.31, the strong effect is due to a smaller a distance between the light frequency and the Mie-resonance. With decreasing frequency offset due to a different dielectric function, an increase of interference effects is observed from roughly 4.8 % up to ca. 10.5 %. The effects originating from interference phenomena are therefore more than doubled.*

5.8 Conclusion

We presented in this chapter a theoretical and microscopic model of random lasing. Our theory is based on the coupling of the semi-classical laser rate equations describing the atomic population inversion to the microscopic theory of light transport in the presence of optical gain. This transport theory has been developed here for a system of finite size, resembling an infinitely extended film geometry, i.e. finite in one spatial dimension but infinite within the plane of the film.

This model represents the logical combination of the work we presented in chapters 3 and 4. Compared to chapter 3, we took into account the finite size of the system, which provides intensity loss through the surfaces as well as an appropriate theory of laser behavior in terms of the laser rate equation for a four level laser system. In contrast to chapter 4, we now considered the correct microscopic description of light intensity transport in such systems.

Therefore we establish a fully self-consistent treatment, of both the transport characteristics via the self-consistent diffusion coefficient including of course also interference contributions on the one hand, and the fully self-consistent coupling of this intensity transport to laser rate equations.

Of particular interest to us is the stationary laser behavior where the constant and isotropic external pumping and the emitted light intensity establish a balancing. For this case we showed that the corresponding microscopic energy-density correlation function always behaves causal as it must. This fact is due to intensity loss through the samples surface, which therefore stabilizes the lasing mode. This means that photon density which is generated within the sample via pumping and amplification, is balanced by the loss of intensity thru the surfaces of the sample. Therefore the photon density in the sample remains finite at all times. The second major result in this section is the calculation of the correlation function, describing the spot size of the lasing modes. We find that the spot size decreases with increasing pumping according to an inverse square root behavior. This decay of the spot size is in qualitative and even quantitative agreement with experimental data.

Chapter 6

Summary

This thesis is divided in three parts which lead in sum to the description of a Random Laser. First we developed a transport theory for light in disordered media. This theory is based on a description of Vollhardt and Wölfle for electrons and it is valid for infinite translational invariant bulk material. But this description is actually only for certain assumptions valid for the description of a random laser. This point is discussed in the second part of the thesis. Here a diffusive system coupled to four-level laser rate equations is solved self-consistently. Finally a self-consistent transport theory for specimens of finite size is developed and coupled to the laser rate equations. Solving this system self-consistently includes that the intrinsic gain is also calculated self-consistently and not included by an external parameter. All three parts of this thesis give (valid in their regimes) striking evidence to the size of the correlation length and thus the lasing volume, which has been compared to the experimentally determined spot size. We find an inverse square root decay of the correlation length. Additionally we derive a linearly increasing photon number well above the lasing threshold by solving this theory. These results show, that the theory is capable of describing the random laser experiments. Finally our calculations show the importance of the interference effects for the onset of lasing. We see that the Cooperons contribute the more the frequency approaches the Mie-resonances. This is an important remark in the discussion whether localization is or is not important for the onset of random lasing.

Appendix A

Dyson Equations

In 3 dimensional media the following general relations hold

$$
\begin{aligned}
G(\vec{r}, z; \vec{r}\,', z\,'; \omega) &= G_0(\vec{r}, z; \vec{r}\,', z\,'; \omega) \\
&+ \int d\vec{r}_1 \int d\vec{r}_2 G(\vec{r}, z; \vec{r}_2, z_2; \omega) \Sigma(\vec{r}_2, z_2; \vec{r}_3, z_3; \omega) \\
&\times G(\vec{r}_3, z_3; \vec{r}\,', z\,'; \omega)
\end{aligned}
\tag{A.1}
$$

Since all the above functions are averaged quantities, they are translational invariant so that one can write

$$
\begin{aligned}
G(\vec{r} - \vec{r}\,', z - z\,'; \omega) &= G_0(\vec{r} - \vec{r}\,', z - z\,'; \omega) - \\
\int d\vec{r}_1 \int dz_1 \int d\vec{r}_2 &\int dz_2 G(\vec{r} - \vec{r}_1, z - z_1; \omega) \Sigma(\vec{r}_1 - \vec{r}_2, z_1 z_2; \omega) \\
&\times G(\vec{r}_2 - \vec{r}\,', z_2 - z\,'; \omega)
\end{aligned}
\tag{A.2}
$$

Furthermore, usually one assumes a local selfenergy, i.e. $\Sigma(\vec{r}_1 - \vec{r}_2, z_1 z_2; \omega) = \Sigma(\omega)\delta(\vec{r}_1 - \vec{r}_2)\delta(z_1 - z_2)$ which is then momentum independent in momentum space:

$$
\begin{aligned}
G(\vec{r} - \vec{r}\,', z, z\,'; \omega) &= G_0(\vec{r} - \vec{r}\,', z, z\,'; \omega) + \\
\int d\vec{r}_1 \int dz_1 \int d\vec{r}_2 &\int dz_2 G(\vec{r} - \vec{r}_1, z, z_1; \omega) \\
&\times \Sigma(\omega)\delta(\vec{r}_1 - \vec{r}_2)\delta(z_1 - z_2) G(\vec{r}_2 - \vec{r}\,', z_2, z\,'; \omega)
\end{aligned}
\tag{A.3}
$$

Therefore the integrals regarding $\vec{r}_1$ and z_1 can be solved resulting in

$$
\begin{aligned}
G(\vec{r} - \vec{r}\,', z - z\,'; \omega) &= G_0(\vec{r} - \vec{r}\,', z - z\,'; \omega) + \\
\int d\vec{r}_2 \int dz_2 &G(\vec{r} - \vec{r}_2, z - z_2; \omega) \Sigma(\vec{r}_2, z_2) G(\vec{r}_2 - \vec{r}\,', z_2 - z\,'; \omega)
\end{aligned}
\tag{A.4}
$$

For the not averaged Green's function, we obtain from the wave equation the following relation

$$
\begin{aligned}
\tilde{G}(\vec{r}, z; \vec{r}\,', z\,'; \omega) &= \tilde{G}_0(\vec{r}, z; \vec{r}\,', z\,'; \omega) + \\
\int d\vec{r}_2 \int dz_2 &\tilde{G}_0(\vec{r}, z; \vec{r}_2, z_2; \omega) \sigma(\vec{r}_2, z_2; \omega) \tilde{G}(\vec{r}_2, z_2; \vec{r}\,', z\,'; \omega)
\end{aligned}
\tag{A.5}
$$

where the shorthand notation for the "scattering potential" $\sigma(\vec{r}_2, z_2; \omega)$ has been introduced according to

$$
\sigma(\vec{r}_2, z_2; \omega) = -(\epsilon_0 - \epsilon(\vec{r}_2, z_2))\frac{\omega^2}{c^2}
\tag{A.6}
$$

Taking the disorder average of the above Dyson equation, Eq. (A.5) we arrive at

$$\langle \tilde{G}(\vec{r}, z; \vec{r}', z'; \omega)\rangle = \tilde{G}_0(\vec{r}, z; \vec{r}', z'; \omega) + \tag{A.7}$$
$$\int d r_2 \int dz_2 \tilde{G}_0(\vec{r}, z; \vec{r}_2, z_2; \omega)\langle \sigma(\vec{r}_2, z_2; \omega) \tilde{G}(\vec{r}_2, z_2; \vec{r}', z'; \omega)\rangle$$

where $\langle \ldots \rangle$ denotes the disorder average. Comparing this result, Eq. (A.7), with the above Dyson equation for disorder averaged Green's function G, Eq. (A.4), we obtain the relation

$$\langle \sigma(\vec{r}_2, z_2; \omega)\tilde{G}(\vec{r}_2, z_2; \vec{r}', z'; \omega)\rangle = \int dr_1 \Sigma(\vec{r}_2, z_2; \vec{r}_1, z_1; \omega) G(\vec{r}_1, z_1; \vec{r}', z'; \omega) \tag{A.8}$$

Now let us return to the wave equation for the not disorder-averaged Green's function for the particular film-geometry, i.e. from here on the translational invariance in z-direction is explicitly broken

$$(\partial_x^2 + \partial_y^2 + \partial_z^2 + \frac{\epsilon_b}{c^2}\omega^2 - \sigma(\vec{r}, z))\tilde{G}(\vec{r}, z; \vec{r}', z'; \omega) = \delta(\vec{r} - \vec{r}')\delta(z - z'). \tag{A.9}$$

Taking here the disorder average of Eq. (A.9), we obtain

$$(\partial_x^2 + \partial_y^2 + \partial_z^2 + \frac{\epsilon_b}{c^2}\omega^2)\langle \tilde{G}(\vec{r}, z, \vec{r}', z'; \omega)\rangle - \langle \sigma(\vec{r}, z)\tilde{G}(\vec{r}, z; \vec{r}', z'; \omega)\rangle$$
$$= \delta(\vec{r} - \vec{r}')\delta(z - z') \tag{A.10}$$

With the above identification as derived in Eq. (A.8), this may be written as

$$(\partial_x^2 + \partial_y^2 + \partial_z^2 + \frac{\epsilon_b}{c^2}\omega^2)G(\vec{r} - \vec{r}', z, z'; \omega)$$
$$- \int dr_1 \int dz_1 \Sigma(\vec{r}, z; \vec{r}_1, z_1; \omega) G(\vec{r}_1, z_1; \vec{r}', z'; \omega)$$
$$= \delta(\vec{r} - \vec{r}')\delta(z - z') \tag{A.11}$$

where we used $G(\vec{r} - \vec{r}', z, z'; \omega) \equiv \langle \tilde{G}(\vec{r}, z, \vec{r}', z'; \omega)\rangle$. Now we use again the local approximation for the self-energy, i.e.

$$\Sigma(\vec{r}, z; \vec{r}_1, z_1; \omega) = \Sigma(\omega) \cdot \delta(\vec{r} - \vec{r}_1)\delta(z - z_1) \tag{A.12}$$

and take the Fourier transform of Eq. (A.11), we eventually arrive at

$$(\frac{\epsilon_b}{c^2}\omega^2 - q^2 + \partial_z^2)G(\vec{q}, z, z'; \omega) - \Sigma(\omega)G(\vec{q}, z, z'; \omega) = \delta(z - z') \tag{A.13}$$

or equivalently at the expression

$$(\frac{\epsilon_b}{c^2}\omega^2 - q^2 - \Sigma(\omega) + \partial_z^2)G(\vec{q}, z, z'; \omega) = \delta(z - z') \tag{A.14}$$

This equation, Eq. (A.14), may be used to define an operator, which is the inverse of $(\frac{\epsilon_b}{c^2}\omega^2 - q^2 - \Sigma(\omega) + \partial_z^2)$. This operator defined as $\hat{G}(\vec{q}, z, z'; \omega)$ may then be written as

$$\hat{G}(\vec{q}, z, z'; \omega) = \frac{\delta(z - z')}{\frac{\epsilon_b}{c^2}\omega^2 - q^2 - \Sigma(\omega) + \partial_z^2} \tag{A.15}$$

Since the context in which the Green's function operator appears is always uniquely determining whether $\hat{G}(\vec{q}, z, z'; \omega)$, Eq. (A.15), or rather $G(\vec{q}, z, z'; \omega)$, Eq. (A.14), is meant, we subsequently will use the notation $G(\vec{q}, z, z'; \omega) = \hat{G}(\vec{q}, z, z'; \omega)$.

The position space representation of the above Operator $G(\vec{q}, z, z'; \omega)$, Eq. (A.15), may analytically be found by explictly solving the above differential equation, Eq. (A.14). The solution is presented below.

Appendix B

Disorder Averaged Full Single-Particle Green's Function

The free Green's function $G_0(\vec{r}, z; \omega)$ is defined by

$$(\partial_x^2 + \partial_y^2 - \partial_z^2 + \frac{\epsilon_b}{c^2}\omega^2)G_0(\vec{r}, z; \vec{r}', z'; \omega) = \delta(\vec{r} - \vec{r}')\delta(z - z') \tag{B.1}$$

which can be written

$$(-q^2 + \partial_z^2 + \frac{\epsilon_b}{c^2}\omega^2)G_0(\vec{q}, z; z'; \omega) = \delta(z - z') \tag{B.2}$$

Also relevant to us is $G_0(\vec{q}, z, z; \omega)$ obeying

$$(-q^2 + \partial_z^2 + \frac{\epsilon_b}{c^2}\omega^2)G_0(\vec{q}, z, z; \omega) = 1 \tag{B.3}$$

Therefore we can write for the operator $G_0^{-1}(\vec{q}, z, z; \omega)$

$$G_0^{-1}(\vec{q}, z; \omega) = (\frac{\epsilon_b}{c^2}\omega^2 - q^2 + \partial_z^2) \tag{B.4}$$

On the other hand Eq. (B.3) can be written as

$$\partial_z^2 G_0(\vec{q}, z, z; \omega) = 1 - (\frac{\epsilon_b}{c^2}\omega^2 - q^2)G_0(\vec{q}, z, z; \omega) \tag{B.5}$$

This differential equation has the general solution

$$\begin{aligned}G_0(\vec{q}, z, z; \omega) &= \frac{1}{\frac{\epsilon_b}{c^2}\omega^2 - q^2 + i0} \\ &+ c_1 \sin\left(z\sqrt{\frac{\epsilon_b}{c^2}\omega^2 - q^2}\right) \\ &+ c_2 \cos\left(z\sqrt{\frac{\epsilon_b}{c^2}\omega^2 - q^2}\right)\end{aligned} \tag{B.6}$$

In Eq. (B.5) the coefficient of G_0 on the r.h.s. is independent of z, i.e. the curvature is fixed, hence it follows that as boundary conditions only the value of G_0 at the boundaries $z = \pm d/2$ can be chosen (Dirichlet boundary conditions). In the following we will always consider symmetric conditions w.r.t. z, therefore we require

$$G_0(\vec{q}, z = \pm d/2; \omega) = R \qquad (B.7)$$

where the value of R will be specified later. With such boundary conditions, the Green's function $G_0(\vec{q}, z; \omega)$ is calculated to be

$$G_0(\vec{q}, z, z; \omega) = \frac{1}{\frac{\epsilon_b}{c^2}\omega^2 - q^2 + i\mathcal{O}}$$
$$+ \left(R - \frac{1}{\frac{\epsilon_b}{c^2}\omega^2 - q^2 + i\mathcal{O}}\right) \frac{\cos\left(z\sqrt{\frac{\epsilon_b}{c^2}\omega^2 - q^2}\right)}{\cos\left(d/2 \cdot \sqrt{\frac{\epsilon_b}{c^2}\omega^2 - q^2}\right)} \qquad (B.8)$$

The same line of arguments applies also to the case of the disorder averaged full Green's function $G(\vec{q}, z; \omega)$, which contains the selfenergy $\Sigma(\omega)$. The detailed discussion regarding the full disorder averaged Green's function is presented below and justifies the following.
In conlusion, we find for the disorder-averaged single particle Green's functions the following expressions

$$G_0^{-1}(\vec{q}, z, z; \omega) = \left(\frac{\epsilon_b}{c^2}\omega^2 - q^2 + \partial_z^2\right) \qquad (B.9)$$
$$G^{-1}(\vec{q}, z, z; \omega) = \left(\frac{\epsilon_b}{c^2}\omega^2 - q^2 - \Sigma(\omega) + \partial_z^2\right) \qquad (B.10)$$

and for z−coordinate space representation of these Green's function operators, we find, as discussed in detail above, the following expressions

$$G_0(\vec{q}, z, z; \omega) = \frac{1}{\frac{\epsilon_b}{c^2}\omega^2 - q^2 + i\mathcal{O}} \qquad (B.11)$$
$$+ \left(R - \frac{1}{\frac{\epsilon_b}{c^2}\omega^2 - q^2 + i\mathcal{O}}\right) \frac{\cos\left(z\sqrt{\frac{\epsilon_b}{c^2}\omega^2 - q^2}\right)}{\cos\left(d/2 \cdot \sqrt{\frac{\epsilon_b}{c^2}\omega^2 - q^2}\right)}$$

as well as

$$G(\vec{q}, z, z; \omega) = \frac{1}{\frac{\epsilon_b}{c^2}\omega^2 - q^2 - \Sigma(\omega)} \qquad (B.12)$$
$$+ \left(R - \frac{1}{\frac{\epsilon_b}{c^2}\omega^2 - q^2 - \Sigma(\omega)}\right) \frac{\cos\left(z\sqrt{\frac{\epsilon_b}{c^2}\omega^2 - q^2 - \Sigma(\omega)}\right)}{\cos\left(d/2 \cdot \sqrt{\frac{\epsilon_b}{c^2}\omega^2 - q^2 - \Sigma(\omega)}\right)}$$

Please note that $G(\vec{q}, z, z; \omega)$ determines the local density of states $N(\omega, z)$, the so-called LDOS, of the system via the relation

$$N(\omega, z) = \omega \mathrm{Im} \int \frac{d^2\vec{q}}{(2\pi)^2} G(\vec{q}, z, z; \omega). \qquad (B.13)$$

B.1 Analytic Calculation of $G(\vec{r}, \vec{r}\,')$

This section is concerned with the analytical calculation of the disorder averaged full single-particle Green's function $G(\vec{q}, z, z'; \omega)$. As seen above the position space representation of the full Green's function is obtained by solving the equation

$$\left(\frac{\epsilon_b}{c^2}\omega^2 - q^2 + \partial_z^2\right) G(\vec{q}, z, z'; \omega) - \Sigma(\omega) G(\vec{q}, z, z'; \omega) = \delta(z - z') \qquad (B.14)$$

Rewriting

$$\left(\frac{\epsilon_b}{c^2}\omega^2 - q^2 - \Sigma(\omega)\right) G(\vec{q}, z, z'; \omega) + \partial_z^2 G(\vec{q}, z, z'; \omega) = \delta(z - z') \qquad (B.15)$$

B.1. ANALYTIC CALCULATION OF $G(\vec{R},\vec{R}')$

$$\partial_z^2 G(\vec{q},z,z';\omega) = \delta(z-z') - \left(\frac{\epsilon_b}{c^2}\omega^2 - q^2 - \Sigma(\omega)\right) G(\vec{q},z,z';\omega), \tag{B.16}$$

we recognize this as a differential equation, which is to be solved using boundary conditions of the Dirichlet type, i.e. we have to specify the following (symmetric) boundary conditions

$$G(z=+\frac{d}{2}) = R \qquad G(z=-\frac{d}{2}) = R \tag{B.17}$$

where d is the width of the sample and the explicit value of R, the value of the Green's function at the surface, will be specified later.

We start the solution by first considering the case $z \neq z'$, where the delta-function does not contribute. There, the differential equation reduces to

$$\partial_z^2 G(\vec{q},z,z';\omega) + \left[\frac{\epsilon_b}{c^2}\omega^2 - q^2 - \Sigma(\omega)\right] G(\vec{q},z,z';\omega) = 0. \tag{B.18}$$

The solution is obviously given by

$$G(\vec{q},z,z';\omega) = \quad A\cos\left(\sqrt{\left[\frac{\epsilon_b}{c^2}\omega^2 - q^2 - \Sigma(\omega)\right]} z\right)$$

$$+ B\sin\left(\sqrt{\left[\frac{\epsilon_b}{c^2}\omega^2 - q^2 - \Sigma(\omega)\right]} z\right) \tag{B.19}$$

with coefficients A and B to be determinded by the boundary conditions. These coefficients will, of course, be functions of (z',q,ω), i.e.

$$A \equiv A(z',q,\omega) \qquad \text{and} \qquad B \equiv B(z',q,\omega). \tag{B.20}$$

The two regimes with $z < z'$ and $z > z'$ will be treated separately, the reason for this will become clear below. Strictly speaking we therefore start with an ansatz

$$G(\vec{q},z,z';\omega) = \begin{cases} C_1\cos\left(\sqrt{a}z\right) + C_2\sin\left(\sqrt{a}z\right) & z < z' \\ C_3\cos\left(\sqrt{a}z\right) + C_4\sin\left(\sqrt{a}z\right) & z > z' \end{cases} \tag{B.21}$$

where the abbreviation $a = \left[\frac{\epsilon_b}{c^2}\omega^2 - q^2 - \Sigma(\omega)\right]$ has been introduced together with the coefficients, still to be determined, C_1, C_2, C_3 and C_4.

For $z < z'$ we find the boundary condition at the lower bound $z = -\frac{d}{2}$

$$G(\vec{q},z=-\frac{d}{2},z';\omega) = \quad C_1\cos\left(\sqrt{\left[\frac{\epsilon_b}{c^2}\omega^2 - q^2 - \Sigma(\omega)\right]}(-\frac{d}{2})\right)$$

$$+ C_2\sin\left(\sqrt{\left[\frac{\epsilon_b}{c^2}\omega^2 - q^2 - \Sigma(\omega)\right]}(-\frac{d}{2})\right) \tag{B.22}$$

$$= R \tag{B.23}$$

or in a clearer form

$$R = \quad C_1\cos\left(\sqrt{\left[\frac{\epsilon_b}{c^2}\omega^2 - q^2 - \Sigma(\omega)\right]}\frac{d}{2}\right)$$

$$- C_2\sin\left(\sqrt{\left[\frac{\epsilon_b}{c^2}\omega^2 - q^2 - \Sigma(\omega)\right]}\frac{d}{2}\right). \tag{B.24}$$

The case $G(\vec{q},z=\frac{d}{2},z';\omega) = R$ is not to be considered, since we required $z < z'$ and for $z = \frac{d}{2}$, i.e. z at the upper boundary, z' will not be within the allowed interval of z-values.

Considering the second regime $z > z'$, we find the boundary condition at upper bound

$$G(\vec{q},z=+\frac{d}{2},z';\omega) = \quad C_3\cos\left(\sqrt{\left[\frac{\epsilon_b}{c^2}\omega^2 - q^2 - \Sigma(\omega)\right]}\frac{d}{2}\right)$$

$$+ C_4\sin\left(\sqrt{\left[\frac{\epsilon_b}{c^2}\omega^2 - q^2 - \Sigma(\omega)\right]}\frac{d}{2}\right) \tag{B.25}$$

$$= R \tag{B.26}$$

or again rewritten

$$R = C_3 \cos\left(\sqrt{\left[\frac{\epsilon_b}{c^2}\omega^2 - q^2 - \Sigma(\omega)\right]}\frac{d}{2}\right)$$
$$+ C_4 \sin\left(\sqrt{\left[\frac{\epsilon_b}{c^2}\omega^2 - q^2 - \Sigma(\omega)\right]}\frac{d}{2}\right) \quad \text{(B.27)}$$

The case $G(\vec{q}, z = -\frac{d}{2}, z'; \omega) = R$ is again not to be considered, since we required $z > z'$ and for $z = -\frac{d}{2}$, i.e. z at the lower boundary, z' will not be within allowed interval of z-values. This is in complete analogy to the above.

So far, we have two equations, Eq. (B.24) and Eq. (B.27), for the four unknown coefficients C_1, C_2, C_3 and C_4.

The third equation is obtained by requiring that the Green's function ought to be continuously at $z = z'$, i.e. we have to require

$$C_3 \cos\left(\sqrt{\left[\frac{\epsilon_b}{c^2}\omega^2 - q^2 - \Sigma(\omega)\right]}z'\right) + C_4 \sin\left(\sqrt{\left[\frac{\epsilon_b}{c^2}\omega^2 - q^2 - \Sigma(\omega)\right]}z'\right)$$
$$=$$
$$C_1 \cos\left(\sqrt{\left[\frac{\epsilon_b}{c^2}\omega^2 - q^2 - \Sigma(\omega)\right]}z'\right) + C_2 \sin\left(\sqrt{\left[\frac{\epsilon_b}{c^2}\omega^2 - q^2 - \Sigma(\omega)\right]}z'\right) \quad \text{(B.28)}$$

The fourth and final equation regards the discontinuity of the first derivative of $G(\vec{q}, z, z'; \omega)$. To find this condition we go back to the defining differential equation in Eq. (B.16) and integrate this equation over z from $z' - \epsilon$ to $z' + \epsilon$, where ϵ is a small positive quantity, and taking the limit $\epsilon \to 0$

$$\lim_{\epsilon \to 0} \int_{z'-\epsilon}^{z'+\epsilon} dz \partial_z^2 G(\vec{q}, z, z'; \omega) = \lim_{\epsilon \to 0} \int_{z'-\epsilon}^{z'+\epsilon} dz \delta(z - z') \quad \text{(B.29)}$$
$$- \lim_{\epsilon \to 0} \int_{z'-\epsilon}^{z'+\epsilon} dz \left(\frac{\epsilon_b}{c^2}\omega^2 - q^2 - \Sigma(\omega)\right) G(\vec{q}, z, z'; \omega),$$

which yields

$$\partial_z G(\vec{q}, z' + \epsilon, z'; \omega) - \partial_z G(\vec{q}, z' - \epsilon, z'; \omega) = 1 \quad \text{(B.30)}$$

because the second term on the r.h.s. of Eq. (B.29) vanishes in the limit $\epsilon \to 0$ since it becomes an integral over a single point. The first term on the l.h.s of Eq. (B.30) clearly refers to the regime $z > z'$, where as the second term on the l.h.s. refers to the case $z < z'$. Therefore Eq. (B.30) constitutes the fourth condition on the coefficients, which reads

$$-\sqrt{\left[\frac{\epsilon_b}{c^2}\omega^2 - q^2 - \Sigma(\omega)\right]}C_3 \sin\left(\sqrt{\left[\frac{\epsilon_b}{c^2}\omega^2 - q^2 - \Sigma(\omega)\right]}z'\right)$$
$$+\sqrt{\left[\frac{\epsilon_b}{c^2}\omega^2 - q^2 - \Sigma(\omega)\right]}C_4 \cos\left(\sqrt{\left[\frac{\epsilon_b}{c^2}\omega^2 - q^2 - \Sigma(\omega)\right]}z'\right)$$
$$- \{-\sqrt{\left[\frac{\epsilon_b}{c^2}\omega^2 - q^2 - \Sigma(\omega)\right]}C_1 \sin\left(\sqrt{\left[\frac{\epsilon_b}{c^2}\omega^2 - q^2 - \Sigma(\omega)\right]}z'\right)$$
$$+\sqrt{\left[\frac{\epsilon_b}{c^2}\omega^2 - q^2 - \Sigma(\omega)\right]}C_2 \cos\left(\sqrt{\left[\frac{\epsilon_b}{c^2}\omega^2 - q^2 - \Sigma(\omega)\right]}z'\right)\}$$
$$= 1 \quad \text{(B.31)}$$

or equivalently the fourth condition on the four coefficients reads

B.1. ANALYTIC CALCULATION OF $G(\vec{R},\vec{R}')$

$$-C_3 \sin\left(\sqrt{\left[\frac{\epsilon_b}{c^2}\omega^2 - q^2 - \Sigma(\omega)\right]}z'\right) + C_4 \cos\left(\sqrt{\left[\frac{\epsilon_b}{c^2}\omega^2 - q^2 - \Sigma(\omega)\right]}z'\right)$$
$$+C_1 \sin\left(\sqrt{\left[\frac{\epsilon_b}{c^2}\omega^2 - q^2 - \Sigma(\omega)\right]}z'\right) - C_2 \cos\left(\sqrt{\left[\frac{\epsilon_b}{c^2}\omega^2 - q^2 - \Sigma(\omega)\right]}z'\right)$$
$$= \frac{1}{\sqrt{\left[\frac{\epsilon_b}{c^2}\omega^2 - q^2 - \Sigma(\omega)\right]}} \quad (B.32)$$

Hence the four coefficients $C_1(z',q,\omega), C_2(z',q,\omega), C_3(z',q,\omega)$ and $C_4(z',q,\omega)$ are related by the four equations Eqs. (B.24), (B.27), (B.28) and (B.32). The rather lengthy solution of this system of equations can be done and yields the following results.

$$C_1(z',q,\omega) = \frac{2\sqrt{a}R + \sin\left(\sqrt{a}\left(z' - \frac{d}{2}\right)\right)}{2\sqrt{a}\cos\left(\sqrt{a}\frac{d}{2}\right)} \quad (B.33)$$

$$C_2(z',q,\omega) = \frac{\sin\left(\sqrt{a}\left(z' - \frac{d}{2}\right)\right)}{2\sqrt{a}\sin\left(\sqrt{a}\frac{d}{2}\right)} \quad (B.34)$$

$$C_3(z',q,\omega) = \frac{2\sqrt{a}R - \sin\left(\sqrt{a}\left(z' + \frac{d}{2}\right)\right)}{2\sqrt{a}\cos\left(\sqrt{a}\frac{d}{2}\right)} \quad (B.35)$$

$$C_4(z',q,\omega) = \frac{\sin\left(\sqrt{a}\left(z' + \frac{d}{2}\right)\right)}{2\sqrt{a}\sin\left(\sqrt{a}\frac{d}{2}\right)} \quad (B.36)$$

With the above calculated coefficients $C_1(z',q,\omega), C_2(z',q,\omega), C_3(z',q,\omega)$ and $C_4(z',q,\omega)$ the full Green's function for light waves propagating within the disordered slab geometry is finally determined as

$$G(\vec{q},z,z';\omega) = \begin{cases} C_1(z',q,\omega)\cos(\sqrt{a}z) + C_2(z',q,\omega)\sin(\sqrt{a}z) & z < z' \\ C_3(z',q,\omega)\cos(\sqrt{a}z) + C_4(z',q,\omega)\sin(\sqrt{a}z) & z > z' \end{cases}$$
(B.37)

with coefficients as shown in Eqs. (B.33-B.36), furthermore we used again the shorthand notation $a = \left[\frac{\epsilon_b}{c^2}\omega^2 - q^2 - \Sigma(\omega)\right]$.

Appendix C

Two Particle Green's Function - The Intensity Correlator

Here we want to consider the expression for the intensity $\langle EE^*\rangle$, expressed in terms of single-particle Green's functions, where $\langle \ldots \rangle$ refers to the disorder average. This uses the relation between the electric field and the Green's function as given by

$$E(\vec{r}, z; \omega) = \int d\vec{r}' \int dz' G(\vec{r}, z; \vec{r}', z'; \omega) j(\vec{r}', z'; \omega). \tag{C.1}$$

where the current $j(\vec{r}', z'; \omega)$ represents the source of the field E as described by Maxwell's equations. Therefore the intensity correlation $\langle EE^*\rangle$ reads

$$\langle E(\vec{r}, z; \omega) E^*(\vec{r}', z'; \omega') \rangle = \tag{C.2}$$
$$\int d\vec{r}'' \int dz'' \int d\vec{r}''' \int dz''' \langle \tilde{G}(\vec{r}, z; \vec{r}'', z''; \omega) \tilde{G}^*(\vec{r}', z'; \vec{r}''', z'''; \omega')$$
$$\times j(\vec{r}'', z''; \omega) j^*(\vec{r}''', z'''; \omega') \rangle$$

and finally by utilizing the fact that the sources do not depend on the particular disorder realization

$$\langle E(\vec{r}, z; \omega) E^*(\vec{r}', z'; \omega') \rangle = \tag{C.3}$$
$$\int d\vec{r}'' \int dz'' \int d\vec{r}''' \int dz''' \langle \tilde{G}(\vec{r}, z; \vec{r}'', z''; \omega) \tilde{G}^*(\vec{r}', z'; \vec{r}''', z'''; \omega') \rangle$$
$$\times j(\vec{r}'', z''; \omega) j^*(\vec{r}''', z'''; \omega')$$

From these considerations it becomes clear to define the disorder averaged two-particle Green's function Γ as

$$\Gamma(\vec{r}, \vec{r}'', \vec{r}', \vec{r}'''; z, z'', z', z'''; \omega, \omega') := \langle \tilde{G}(\vec{r}, z; \vec{r}'', z''; \omega) \tilde{G}^*(\vec{r}', z'; \vec{r}''', z'''; \omega') \rangle \tag{C.4}$$

For later use, we calculate at this point the Fourier transform of Γ, where we use the partial Fourier transform within the (x, y)-plane only, as defined in above

$$\Gamma(\vec{r}, \vec{r}'', \vec{r}', \vec{r}'''; z, z'', z', z'''; \omega, \omega') = \int \frac{d^2q}{(2\pi)^2} \int \frac{d^2q''}{(2\pi)^2} \int \frac{d^2q'}{(2\pi)^2} \int \frac{d^2q'''}{(2\pi)^2}$$
$$\times e^{+i\vec{q}\cdot\vec{r}} e^{-i\vec{q}''\cdot\vec{r}''} e^{-i\vec{q}'\cdot\vec{r}'} e^{+i\vec{q}'''\cdot\vec{r}'''}$$
$$\times \langle \tilde{G}(\vec{q}, z; \vec{q}'', z''; \omega) \tilde{G}^*(\vec{q}', z'; \vec{q}''', z'''; \omega') \rangle \tag{C.5}$$

where we identify

$$\Gamma(\vec{q},\vec{q}'',\vec{q}',\vec{q}''';z,z'',z',z''';\omega,\omega') = \langle \tilde{G}(\vec{q},z;\vec{q}'',z'';\omega)\tilde{G}^*(\vec{q}',z';\vec{q}''',z''';\omega)\rangle \tag{C.6}$$

Observing that the disorder averaged quantity within the (x,y)-plane is clearly translational invariant, we can further write, by extracting the corresponding delta function,

$$\Gamma(\vec{q},\vec{q}'',\vec{q}',\vec{q}''';z,z'',z',z''';\omega,\omega') = \tag{C.7}$$
$$(2\pi)^2\delta(\vec{q}-\vec{q}''-\vec{q}'+\vec{q}''')\Gamma(\vec{q},\vec{q}'',\vec{q}',\vec{q}''';z,z'',z',z''';\omega,\omega')$$

From this result it becomes advisable to introduce new momenta by the relations

$$\vec{Q} := \vec{q}-\vec{q}' \tag{C.8}$$
$$\vec{Q}' := \vec{q}''-\vec{q}''' \tag{C.9}$$
$$\vec{p} := \frac{1}{2}(\vec{q}+\vec{q}') \tag{C.10}$$
$$\vec{p}' := \frac{1}{2}(\vec{q}''+\vec{q}''') \tag{C.11}$$

The inverse transformation is then given by

$$\vec{q} = \vec{p}+\frac{1}{2}\vec{Q} \tag{C.12}$$
$$\vec{q}' = \vec{p}-\frac{1}{2}\vec{Q} \tag{C.13}$$
$$\vec{q}'' = \vec{p}'+\frac{1}{2}\vec{Q}' \tag{C.14}$$
$$\vec{q}''' = \vec{p}'-\frac{1}{2}\vec{Q}' \tag{C.15}$$

By using the above transformation given in Eqs. (C.8)-(C.11), we can rewrite Eq. (C.3)

$$\langle E(\vec{r},z;\omega)E^*(\vec{r}',z';\omega')\rangle = \tag{C.16}$$
$$\int d r'' \int dz'' \int d r''' \int dz''' \int \frac{d^2p}{(2\pi)^2} \int \frac{d^2p'}{(2\pi)^2} \int \frac{d^2Q}{(2\pi)^2} \int \frac{d^2Q'}{(2\pi)^2}$$
$$\times e^{+i(\vec{p}+\vec{Q}/2)\cdot\vec{r}} e^{-i(\vec{p}'+\vec{Q}'/2)\cdot\vec{r}''} e^{-i(\vec{p}-\vec{Q}/2)\cdot\vec{r}'} e^{+i(\vec{p}'-\vec{Q}'/2)\cdot\vec{r}'''}$$
$$\times (2\pi)^2\delta(\vec{q}-\vec{q}''-\vec{q}'+\vec{q}''')\Gamma(\vec{q},\vec{q}'',\vec{q}',\vec{q}''';z,z'',z',z''';\omega,\omega')$$
$$\times \delta(\vec{r}''-\vec{r}''')\delta(z''-z''')J(\vec{r}'',z'') \tag{C.17}$$

where we have used

$$j(\vec{r}'',z'';\omega)j^*(\vec{r}''',z''';\omega') = \delta(\vec{r}''-\vec{r}''')\delta(z''-z''')J(\vec{r}'',z''). \tag{C.18}$$

By integration we obtain

$$\langle E(\vec{r},z;\omega)\; E^*(\vec{r}',z';\omega')\rangle = \int dz'' \int \frac{d^2p}{(2\pi)^2} \int \frac{d^2p'}{(2\pi)^2} \int \frac{d^2Q}{(2\pi)^2} \tag{C.19}$$
$$\times e^{+i\vec{p}\cdot(\vec{r}-\vec{r}')} e^{-i\vec{Q}/2\cdot(\vec{r}+\vec{r}')} \Gamma_{pp'}(\vec{Q}''';z,z'',z',z'';\omega,\omega') J(\vec{Q},z'')$$

In analogy to the transformation in momentum space shown in Eqs. (C.8)-(C.11), we now define a similar transformation with respect to the
temporal coordinates, i.e. the frequencies according to

$$\tilde{\omega} := \frac{1}{2}(\omega + \omega') \tag{C.20}$$

$$\Omega := (\omega - \omega') \tag{C.21}$$

with the corresponding inverse transformation

$$\omega = \tilde{\omega} + \frac{1}{2}\Omega \tag{C.22}$$

$$\omega' = \tilde{\omega} - \frac{1}{2}\Omega \tag{C.23}$$

Using the above defined transformation we may eventually find

$$\langle E(\vec{r}, z\; \omega + \frac{\Omega}{2}) E^*(\vec{r}', u; \omega - \frac{\Omega}{2}) \rangle = \tag{C.24}$$
$$\int dz'' \int \frac{d^2p}{(2\pi)^2} \int \frac{d^2p'}{(2\pi)^2} \int \frac{d^2Q}{(2\pi)^2} e^{+i\vec{p}\cdot(\vec{r}-\vec{r}')} e^{-i\vec{Q}/2\cdot(\vec{r}+\vec{r}')}$$
$$\times \Gamma^\omega_{pp'}(\vec{Q}; z, z''; u, z''; \Omega) J(\vec{Q}, z'')$$

Therefore we will consider in the following only the two-particle Green's function $\Gamma^\omega_{pp'}(\vec{Q}; z, z'; u, u''; \Omega)$. It can be shown, see Appendix C, that this quantity obeys the following equation, the so-called *Bethe-Salpeter equation*

$$\Gamma^\omega_{pp'}(\vec{Q}, \Omega; z, z'; u, u') = G^{\omega+}_{p+}(\vec{Q}, \Omega; z, z') \left(G^{\omega-}_{p-}\right)^*(\vec{Q}, \Omega; u, u')$$
$$\times \left[\delta(\vec{p} - \vec{p}')\right. \tag{C.25}$$
$$+ \int dz'' \int dz''' \int du'' \int du''' \int \frac{d^2p''}{2\pi}$$
$$\times G^{\omega+}_{p+}(\vec{Q}, \Omega; z, z'') \left(G^{\omega-}_{p-}\right)^*(\vec{Q}, \Omega; u, u'') \gamma^\omega_{pp''}(\vec{Q}, \Omega; z'', z'''; u'', u''')$$
$$\left. \times \Gamma^\omega_{p''p'}(\vec{Q}, \Omega; z''', z'; u''', u')\right]$$

introducing the so-called irreducible vertex function $\gamma^\omega_{pp''}(\vec{Q}, \Omega; z'', z'''; u'', u''')$.
Since in all applications the correlator with a *single starting point* appears, we set $z' = u'$ and call this point s in the following. Furthermore, we use now center of mass and relative coordinates for z as well, for example

$$Z := \frac{z + u}{2} \qquad \Delta z := z - u \tag{C.26}$$

$$Z'' := \frac{z'' + u''}{2} \qquad \Delta z'' := z'' - u'' \tag{C.27}$$

$$Z''' := \frac{z''' + u'''}{2} \qquad \Delta z''' := z''' - u''' \tag{C.28}$$

Using the abbreviation $Z_\pm := Z \pm \frac{\Delta z}{2}$ etc., we can eventually write

$$\Gamma^\omega_{pp'}(\vec{Q},\Omega;Z_+,s;Z_-,s) = G^{\omega+}_{p_+}(\vec{Q},\Omega;Z_+,s)\left(G^{\omega-}_{p_-}\right)^*(\vec{Q},\Omega;Z_-,s)$$

$$\times \left[\delta\left(\vec{p}-\vec{p}\,'\right) + \int dZ'' \int d\Delta z'' G^{\omega+}_{p_+}(\vec{Q},\Omega;Z_+,Z''_+)\left(G^{\omega-}_{p_-}\right)^*(\vec{Q},\Omega;Z_-,Z''_-)\right.$$

$$\times \int dZ''' \int d\Delta z''' \int \frac{d^2 p''}{2\pi}$$

$$\left.\times \gamma^\omega_{pp''}(\vec{Q},\Omega;Z''_+,Z'''_+;Z''_-,Z'''_-)\Gamma^\omega_{p''p'}(\vec{Q},\Omega;Z'''_+,s;Z'''_-,s)\right]$$

In a last step we will employ a Fourier transformation with respect to Δz, i.e. we use the transformation

$$\Gamma^\omega_{pp'}(\vec{Q},\Omega;Z,s;Z,s) = \int \Delta z \exp\left(ip_z \Delta z\right)\, \Gamma^\omega_{pp'}(\vec{Q},\Omega;Z_+,s;Z_-,s) \quad (C.29)$$

introducing the relative momentum (component) p_z. This ensures that the above introduced relative momentum vector $\vec{p}$ is from now on again a three-dimensional vector. This Fourier transformation with respect to the relative spatial coordinate even in the z-direction is valid as long as the width d of the considered film-geometry is large compared to the wavelength of light λ, i.e. if

$$d \ll \lambda \quad (C.30)$$

is true. In terms of physics, this means that for quantities which change on the scale of the wavelength, the system is translational invariant even in the finite z-direction. In experiments, this requirement is perfectly valid, since the wavelength of light is about ~ 1 μm and the width of the film is up to ~ 100 μm.

The result of the Fourier transformation is therefore

$$\Gamma^\omega_{\vec{p}\vec{p}'}(\vec{Q},\Omega;Z,s;Z,s) = G^{\omega+}_{\vec{p}_+}(\vec{Q},\Omega;Z,s)\left(G^{\omega-}_{\vec{p}_-}\right)^*(\vec{Q},\Omega;Z,s)$$

$$\times \left[\delta\left(\vec{p}-\vec{p}\,'\right) + \int dZ'' G^{\omega+}_{\vec{p}_+}(\vec{Q},\Omega;Z,Z'')\left(G^{\omega-}_{\vec{p}_-}\right)^*(\vec{Q},\Omega;Z,Z'')\right.$$

$$\times \int dZ''' \int \frac{d^2 p''}{2\pi} \quad (C.31)$$

$$\left.\times \gamma^\omega_{\vec{p}\vec{p}''}(\vec{Q},\Omega;Z'',Z''';Z'',Z''')\Gamma^\omega_{\vec{p}''\vec{p}'}(\vec{Q},\Omega;Z''',s;Z''',s)\right]$$

where all relative momenta, throughout this thesis denoted with lower-case characters, are now three-dimensional vectors, whereas the center of mass coordinate remains Fourier transformed with respect to the (x,y)-plane only.

The above form of the Bethe-Salpeter equation, Eq. (C.31), is the final expression describing light propagation in systems of finite size. The solution of this equation is discussed in the next section.

Appendix D

Calculating the Memory Kernel $M(\Omega)$

In this section we present the calculation of the memory kernel $M(\Omega)$ which apeares in the current-relaxation equation and more importantly in the definition of the full diffusion constant $D(\Omega)$. The memory kernel $M(\Omega)$ contains the full irreducible vertex $\gamma_{p\,p'}$, which was diagrammatically calculated by Vollhardt and Wölfle by its most singular contributions, the so-called Cooperon, the sum of all maximally crossed diagrams, depicted in Fig. D.1a. The Cooperon can be disentangled

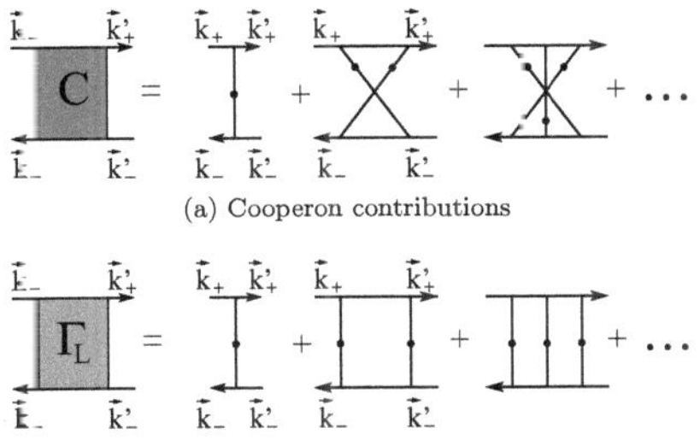

(a) Cooperon contributions

(b) Diffuson contributions

Figure D.1:
(a) The Cooperon **C** *as the sum of the maximally crossed diagrams.*
(b) The particle-hole ladder **Γ** *as sum of all ladder diagrams.*

into a sum of ladder diagrams, as shown in Fig. D.1b, by flipping over one of the propagator lines and exploiting time reversal symmetry. This has been discussed in detail in chapter 2. The result is

$$\mathbf{C}_{p\,p'} = \mathbf{\Gamma}(\vec{\mathbf{q}}) \tag{D.1}$$

with $\vec{\mathbf{q}} = \vec{p} + \vec{p}'$. In any particle-hole diagram which contains a Cooperon, the singularity at $\vec{p} = \vec{p}'$ is strongest if the Cooperon block is crossing the entire diagram diagonally because otherwise the singularity is integrated over and therefore weakened. Therefore, the (infrared) divergent behavior of each diagram contained in the irreducible vertex $\gamma_{p\,p'}$ can be classified by the number of diagonally crossing Cooperons as shown in Fig. D.2. The auxiliary vertex function $\tilde{\gamma}$ is defined as the sum of all those irreducible diagrams which do not contain any diagonally crossing interaction line, and the internal part $\tilde{\Phi}$ is the sum of all reducible and irreducible particle-hole diagrams with the same restriction. By reversing the lower Green's function line in the last diagram in Fig. D.2 one obtains Fig. D.3 which expresses the irreducible vertex in terms of particle-hole ladders, the so-called Diffusons Γ.

By reversing the lower Green's function line in the last diagram in Fig. D.2, the crossing blocks $C_{pp'}$ are disentangled into ladders $\Gamma(\vec{q})$ with the momentum argument $\vec{q}$ replaced by $\vec{q} = \vec{p} + \vec{p}'$, compared to the last diagram in Fig. D.3. From this it is seen, that the internal part $\tilde{\Phi}$ contains all diagrams also present in the energy density correlation function $\Phi_{\rho\rho}$ except for those with ladders Γ at either or both ends, because in the original definition any diagonally crossing interaction lines had been excluded from $\tilde{\Phi}$. Therefore we write

$$\tilde{\Phi}(\vec{q}) = \Phi_{\rho\rho} - R(\vec{q})\Gamma(\vec{q})R(\vec{q}) - R(\vec{q})\Gamma(\vec{q})\tilde{\Phi}(\vec{q}) - \tilde{\Phi}(\vec{q})\Gamma(\vec{q})R(\vec{q})$$
$$- R(\vec{q})\Gamma(\vec{q})\tilde{\Phi}(\vec{q})\Gamma(\vec{q})R(\vec{q}) \tag{D.2}$$

where we defined $R(\vec{q}) = \sum_p G^R_{p-} G^A_{p+}$. Equivalently, we may write

$$\tilde{\Phi}(\vec{q}) \left[1 + 2R(\vec{q})\Gamma(\vec{q}) + R(\vec{q})^2 \Gamma(\vec{q})^2\right] = \Phi_{\rho\rho} - R(\vec{q})\Gamma(\vec{q})R(\vec{q}) \tag{D.3}$$
$$\left[1 + R(\vec{q})\Gamma(\vec{q})\right]^2 \tilde{\Phi}(\vec{q}) = \Phi_{\rho\rho} - R(\vec{q})\Gamma(\vec{q})R(\vec{q}) \tag{D.4}$$

Since the ladder diagrams can be summed up as a geometric series $\Gamma = \frac{\gamma_0}{1 - \gamma_0 R}$ we have $\frac{\Gamma}{\gamma_0} = 1 + R\Gamma$, and can continue to write

$$\tilde{\Phi}(\vec{q}) = \left(\frac{\gamma_0}{\Gamma}\right)^2 \left[\Phi_{\rho\rho} - R(\vec{q})^2 \Gamma(\vec{q})\right] \tag{D.5}$$

From this it can be seen, that the divergencies of the crossing cooperons in the last diagram of Fig. D.2 are cancelled by $\Gamma^{-2}(\vec{q})$ in $\tilde{\Phi}(\vec{q})$ and therefore only the singularity in $\Phi_{\rho\rho}$ itself is left. Additionally diagrams with less than two Cooperon blocks are less divergent and may be neglected and we find

$$\gamma_{pp'} = \Gamma(\vec{q}) + \Gamma(\vec{q})\tilde{\Phi}(\vec{q})\Gamma(\vec{q})$$
$$\gamma_{pp'} = \gamma_0^2 \Phi_{\rho\rho} + \ldots \tag{D.6}$$

Therefore, the irreducible vertex $\gamma_{pp'}$ is dominated by the pole structure of the correlation $\Phi_{\rho\rho}$ and hence by the energy density correlation $\Phi_{\epsilon\epsilon}$, which depends on the diffusion coefficient itself as shown in chapters in 3 and 5 and establishes a selfconsistent equation for the diffusion constant $D(\Omega)$

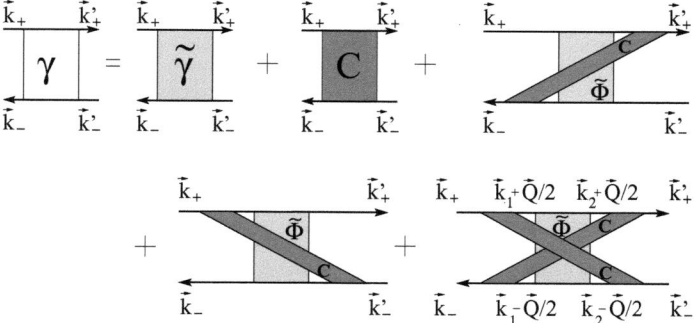

Figure D.2: *Classification of all diagrams of the irreducible vertex in terms of Cooperons.*

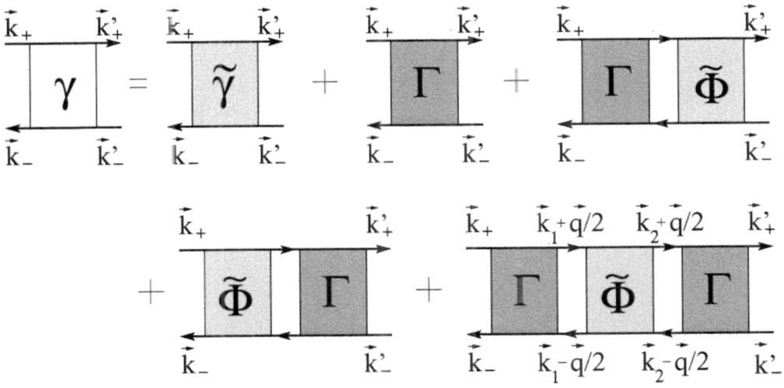

Figure D.3: *Transformation of the diagrams of Fig. D.2, expressing the irreducible vertex in terms of particle-hole ladders, the so-called Diffusons* Γ.

Appendix E

Technical Transformations

E.1 Transformation of the Coordinate System

Here we wish to transform simple position variables to center of mass and relative coordinates. We introduce the following transformation rules:

$$
\begin{aligned}
R &:= \tfrac{r_2+r_1}{2} & \Delta r &:= r_1 - r_2 \\
R' &= \tfrac{r'_2+r'_1}{2} & \Delta r' &:= r'_1 - r'_2 \\
R_{34} &= \tfrac{r_3+r_4}{2} & \Delta r_{34} &:= r_3 - r_4 \\
R_{56} &= \tfrac{r_5+r_6}{2} & \Delta r_{56} &:= r_5 - r_6
\end{aligned}
\tag{E.1}
$$

with inversions given by

$$
\begin{aligned}
r_1 &:= R + \tfrac{\Delta r}{2} & r_2 &:= R - \tfrac{\Delta r}{2} \\
r'_1 &:= R' + \tfrac{\Delta r'}{2} & r'_2 &:= R' - \tfrac{\Delta r'}{2} \\
r_3 &:= R_{34} + \tfrac{\Delta r_{34}}{2} & r_4 &:= R_{34} - \tfrac{\Delta r_{34}}{2} \\
r_5 &:= R_{56} + \tfrac{\Delta r_{56}}{2} & r_6 &:= R_{56} - \tfrac{\Delta r_{56}}{2}
\end{aligned}
\tag{E.2}
$$

According to these rules the derivative $\nabla^2_{r_1}$ changes to

$$\nabla_{r_1} = \tfrac{1}{2}\nabla_R + \nabla_{\Delta r} \tag{E.3}$$

$$\nabla^2_{r_1} = \tfrac{1}{4}\nabla^2_R + \nabla^2_{\Delta r} + \nabla_R \nabla_{\Delta r} \tag{E.4}$$

and analogously for $\nabla^2_{r_2}$

$$\nabla_{r_2} = \tfrac{1}{2}\nabla_R - \nabla_{\Delta r} \tag{E.5}$$

$$\nabla^2_{r_2} = \tfrac{1}{4}\nabla^2_R + \nabla^2_{\Delta r} - \nabla_R \nabla_{\Delta r} \tag{E.6}$$

E.2 Transformation used for Single Particle Greens Function

$$R := \frac{r_1 + r'_1}{2} \qquad \Delta r := r_1 - r'_1 \tag{E.7}$$

$$\tag{E.8}$$

Bibliography

[1] Wiersma, D. S. *The physics and applications of random lasers*. Nature Phys. **4**,359 (2008).

[2] Cao, H. in Waves Random Media **13**, R1 (2003).

[3] Cao, H. et al. *Random laser action in semiconductor powder*. Phys. Rev. Lett. **82**,2278 (1999).

[4] Klingshirn, C. *ZnO: From basics towards applications*. Phys. Stat. Sol. (b) **244**, 3027 (2007).

[5] Koenderink, F., Lagendijk, A. , Vos,W. L. *Optical extinction due to intrinsic structural variations of photonic crystals*. Phys. Rev. B **72**, 153102 (2005).

[6] S. John, *Electromagnetic absorption in a disordered medium near a photon mobility edge*. Phys. Rev. Lett. **53**, 2169 (1984).

[7] P. W. Anderson, *The question of classical localization: A theory of white paint?* Phil. Mag. B **52**, 505 (1985).

[8] Lagendijk, A., van Albada, M. P. , van der Mark, M. B. *Localization of light: The quest for the white hole*. Physica A **140**, 183 (1986).

[9] Lee, P. A. , Ramakrishnan, T. V. *Disordered electronic systems*. Rev. Mod. Phys. **57**, 287 (1985).

[10] Abrahams, E., Anderson, P.W., Licciardello, D. C. , Ramakrishnan, T. V. *Scaling theory of localization: Absence of quantum diffusion in two dimensions*. Phys. Rev. Lett. **42**, 673 (1979).

[11] Cao, H. et al. *Random laser action in semiconductor powder*. Phys. Rev. Lett. **82**, 2278 (1999).

[12] Markushev, V. M., Ryzhkov, M. V. , Briskina, C. M. *Characteristic properties of ZnO random laser pumped by nanosecond pulses*. Appl. Phys. B **84**, 333 (2006).

[13] Anderson, P. W. *Absence of diffusion in certain random lattices*. Phys. Rev. **109**, 1492 (1958).

[14] Vanneste, C. , Sebbah, P. *Selective excitation of localized modes in active random media*. Phys. Rev. Lett. **87**, 183903 (2001).

[15] Wiersma, D. S , Bartolini, P., Lagendijk, A. , Righini, R. *Localization of light in a disordered medium*. Nature *390*, 671 (1997).

[16] John, S. *Strong localization of photons in certain disordered dielectric superlattices*. Phys. Rev. Lett. **58**, 2486 (1987).

[17] C. Baravian, F. Caton, J. Dillet, G. Toussaint, and P. Flaud, Phys. Rev. E **76**, 011409 (2007)

[18] B.A. van Tiggelen, S.E. Skipetrov, Phys. Rev. E. **73**, 045601 (2006) Rapid Communications

[19] R. Frank, A. Lubatsch, and J. Kroha, Phys. Rev. B **73**, 245107 (2006);

[20] K.Muinonen, Waves in Random Media, **14**, 365 (2004)

[21] A. M. Brodsky, G. T. Mitchell, S. L. Ziegler, and L. W. Burgess, Phys. Rev. E **75**, 046605 (2007)

[22] M. Störzer, C. M. Aegerter, and G. Maret, Phys. Rev. E **73**, 065602 (2006)

[23] M. Störzer, P. Gross, C. M. Aegerter, and G. Maret, Phys. Rev. Lett. **96**, 063904 (2006)

[24] S. Gottardo, R. Sapienza, P.D. Garcia, A. Blanco, D.S. Wiersma, C. Lopez, Nature Photonics **2**, 429 (2008)

[25] L. Florescu and S. John, Phys. Rev. E **69**, 046603 (2004)

[26] X. J. Jiang, Q. Li, C. M. Soukoulis, Phys. Rev. B **59**, R9007 (1999).

[27] K. L. van der Molen, P. Zijlstra, A. Lagendijk, A. P. Mosk, Optics Letters **31**, 1432, (2006)

[28] H. Cao, Waves in Random Media **13**, R1 (2003)

[29] A. Yamilov, X. Wu, X. Liu, R. P. H. Chang, and H. Cao, Phys. Rev. Lett. **96**, 083905 (2006)

[30] A. Yamilov, X. Wu, H. Cao, A. L. Burin, Optics Letters, **30**, 2430 (2005)

[31] S. Chang, A. Taflove, A. Yamilov, A. Burin, H. Cao, Optics Letters, **29**, 917 (2004)

[32] M. Nomura, S. Iwamoto, K. Watanabe, N. Kumagai, Y. Nakata, S. Ishida, and Y. Arakawa, Opt. Express, **14**, 6308 (2006)

[33] L. Florescu, K. Busch and S. John, J. Opt. Soc. Am. B **19**, 2215 (2002)

[34] K. Busch and C. Soukoulis, Phys. Rev. Lett. **75**, 3442 (1995)

[35] J. Kroha, C.M. Soukoulis, and P. Wölfle, Phys. Rev. B **47**, 11093 (1993).

[36] T. Kopp, J. Phys. C **17**, 1897, 1918 (1984)

[37] P.W. Anderson, Phys. Rev. **109**, 1492 (1958)

[38] E. Abrahams P. W. Anderson, D. C. Licciardello, and T. V. Ramakrishnan, Phys. Rev. Lett. **42**, 673 (1979).

[39] D. Vollhardt and P. Wölfle, Phys. Rev. Lett. **45**, 842 (1980); Phys. Rev. Rev. B **22**, 4666 (1980)

[40] D.S. Wiersma , P. Bartolini, A. Lagendijk, and R. Righini, Nature **390**, 671 (1997)

[41] F. Scheffold , R. Lenke, R. Tweer, and G. Maret, Nature **398**, 206 (1999)

[42] D.S. Wiersma , J. Gomez Rivas, P. Bartolini, A. Lagendijk, and R. Righini, Nature **398**, 207 (1999)

[43] J.C.J. Paasschens, T.Sh. Misirpashaev, C.W.J. Beenakker, Phys. Rev. B **54**, 11887 (1996)

[44] H. Cao , J. Y. Xu, D. Z. Zhang, S. H. Chang, S. T. Ho, E. W. Seelig, X. Liu, R. P. H. Chang, Phys. Rev. Lett. **84**, 5584 (2000)

[45] H. Cao , Y. Ling, J. Y. Xu, C. Q. Cao, and P. Kumar, Phys. Rev. Lett. **86**, 4524 (2001)

[46] V.M. Apalkov, M.E. Raikh, B. Shapiro, Phys. Rev. Lett. **89**, 016802 (2002)

[47] S. E. Hodges, M. Munroe, J. Cooper, M. G. Raymer, J. Opt. Soc. Am. B **14**, 191 (1997)

[48] S. W. Wieczorek W. W. Chow, Phys. Rev. A **69**, 033811 (2004)

[49] A. Lubatsch, J. Kroha, K. Busch, Phys. Rev. B **71**, 184201 (2005)

[50] J. Kroha, Physica A **167**, 231 (1990)

[51] Letokhov V S 1968 Generation of light by a scattering medium with negative resonance absorption *Sov. Phys. JETP* **26**, 835

[52] Lawandy N M, Balachandran R M, Gomes A S L, Sauvain E 1994 Laser action in strongly scattering media *Nature* **368**, 436

[53] Cao H, Zhao Y G, Ong H C, Ho S T, Dai J Y, Wu J Y, Chang R P H 1998 Ultraviolet lasing in resonators formed by scattering in semiconductor polycrystalline films *Appl. Phys. Lett.* **73**, 3656
Cao H, Zhao Y G, Ho S T, Seelig E W, Wang Q H, Chang R P H 1999 Random Laser Action in Semiconductor Powder *Phys. Rev. Lett.* **82**, 2278
Cao H, Xu J Y, Zhang D Z, Chang S H, Ho S T, Seelig E W, Liu X, Chang R P H 2000 Spatial Confinement of Laser Light in Active Random Media *Phys. Rev. Lett.* **84**, 5584
Cao H 2003 Lasing in random media *Waves Random Media* **13**, R1

[54] Markushev V M, Zolin V F, Briskina C M 1986 Powder laser *Zh. Prikl. Spektrosk.* **45**, 847

[55] Bahoura M, Morris K J, Noginov M A 2002 Threshold and slope efficiency of $Nd_{0.5}La_{0.5}Al_3(BO_3)_4$ ceramic random laser: effect of the pumped spot size *Opt. Commun.* **201**, 405

[56] Klein S, Cregut O, Gindre D, Boeglin A, Dorkenoo K D 2005 Random laser action in organic film during the photopolymerization process *Opt. Expres* **3**, 5387

[57] Polson R C, Vardeny Z V 2004 Random lasing in human tissues *Appl. Phys. Lett.* **85**, 1289

[58] Anderson P W 1958 Absence of Diffusion in Certain Random Lattices *Phys. Rev.* **109**, 1492

[59] Abrahams E, Anderson P W, Licciardello D C, Ramakrishnan T V 1979 Scaling Theory of Localization: Absence of Quantum Diffusion in Two Dimensions *Phys. Rev. Lett.* **42**, 673

[60] Vollhardt D, Wölfle P 1980 Diagrammatic, self-consistent treatment of the Anderson localization problem in $d <= 2$ dimensions *Phys. Rev. B* **22**, 4666

[61] Lubatsch A, Kroha J, Busch K 2005 Theory of light diffusion in disordered media with linear absorption or gain *Phys. Rev. B* **71**, 184201

[62] Lubatsch A, Frank R, Kroha J 2009 Light Transport in Disordered Systems with Gain/Absorption: A Detailed Analysis *top be submitted*

[63] J. Fallert, R. J. B. Dietz, J. Sartor, D. Schneider, C Klingshirn, H. Kalt Nature Photonics **3**, 279 (2009).

[64] P. W. Anderson, Phys. Rev. **109**, 1492 (1958).

[65] D. Vollhardt, P Wölfle, *Diagrammatic, Selfconsistent Treatment of the Anderson Localization Problem in $d \leq 2$ dimesions*, Phys. Rev. B **22**, 4666 (1980)

[66] J. Jackson *Classical Electrodynamics* (Wiley) (1975)

[67] M. Liu, Phys. J. **12** 49 (2002)

[68] M. P. van Albada, B. A. van Tiggelen, A. Lagendijk, A. Tip, *Speed of propagation of classical waves in strongly scattering media* Phys. Rev. Lett. **66**, 3132 (1991)

[69] A. Martinez, *Statistque de polarisation et effet faraday en diffusion multiple de la lumiere* Ph.D. Thesis (Universite Joseph Fourier, Grenoble, France) (1993)

[70] P. Morse, H. Feshbach *Methods of theoretical Physics* (McGraw-Hill) (1953)

[71] P. Sheng, *Inroduction to wave scattering, Localization, and mesoscopic phenomena* (Academis Press) (1995)

[72] S. John, *Electromagnetic Absorption in a Disordered Medium near a Photon Mobility Edge*, Phys. Rev. Lett. **53**, 2169 (1984)

[73] L. Tsang, A. Ishimaru, J. Opt. Soc. Am. **A 1**, 836 (1984)

[74] M. P. v. Albada, A. Lagendijk, Phys. Rev. Lett. **55**, 2692 (1985)

[75] P. E. Wolf, G. Maret, Phys. Rev. Lett. **55**, 2696 (1985)

[76] J. Kroha, C. M. Soukoulis, P. Wölfle, *Localization of Classical Waves in a random medium: Self-consistent theory*, Phys. Rev. B **47**, 11093 (1993)

[77] J. Kroha, *Diagrammatic Self-consistent Theory of Anderson Localization for the Tight-binding Model* Physica A **167**, 231 (1990)

[78] J. Kroha, T. Kopp, P. Wölfle, *Self-consistent Theory of Anderson Localization for the Tight-binding Model with site-diaogonal disorder* Phys. Rev. B **41**, 888 (1990)

[79] Yu. N. Barabanenkov, L. M. Zurk, M. Yu. Barabanenkov, *Poynting's theorem and electromagnetic wave multiple scattering in dense media near resonance: modified radiative transfer equation* J. Electromagn. Waves. Appl. **9**, 1393 (1995)

[80] D. Livdan, A. A. Lisyansky *Transport properties of waves in absorbing random media with microstructure*
Phys. Rev. B **53**, 14843 (1996)

[81] P.Sheng, Z. Q. Zhang, *Scalar-Wave Localization in a Two-Component Composite*, Phys. Rev. Lett. **57**, 1879 (1986)

[82] E. N. Economou, *Green's Functions in Quantum Physics* Springer Verlag (1983)

[83] F. Mandl, G. Shaw, *Quantum Field Theory* (revised Edition) John Wiley ans Sons (1993)

[84] C. M. Soukoulis, E. N. Economou, G. S. Grest, M. H. Cohen, *Existence of Anderson Localization of Classical Waves in a Random Two-Component Medium*, Phys. Rev. Lett. **62**, 575 (1989)

[85] S. John, *Strong localization of photons in certain disordered dielectric superlattices*, Phys. Rev. Lett. **58**, 2486 (1987)

[86] M. P. van Albada, B. A. van Tiggelen, A. Lagendijk, A. Tip, *Speed of propagation of classical waves in strongly scattering media*, Phys. Rev. Lett. **66**, 3132 (1991)

[87] B. A. van Tiggelen, A. Lagendijk, M. P. van Albada, A. Tip, *Speed of light in random media*, Phys. Rev. B **45**, 12233 (1992)

[88] Yu. N. Barabanenkov, V. D. Ozrin, *Problem of light diffusion in strongly scattering media*, Phys. Rev. Lett. **69**, 1364 (1992)

[89] B. A. van Tiggelen, A. Lagendijk, A. Tip, *Comment on: Problem of light diffusion in strongly scattering media*, Phys. Rev. Lett. **71**, 1284 (1993)

[90] D. Livdan, A. A. Lisyansky, J. Opt. Soc. Am. B **13**, 844 (1996)

[91] Yu. N. Barabanenkov, V. D. Ozrin, *Comment on "Ward identities for transport of classical waves in disordered media"*, Phys. Rev. E. **64**, 18601 (2001)

[92] H. T. Nieh, L. Chen, P. Sheng, *Reply to "Comment on 'Ward identities for transport of classical waves in disordered media' "*, Phys. Rev. E. **64** 18602 (2001)

[93] H. T. Nieh, L. Chen, P. Sheng, *Ward identities for transport of classical waves in disordered media*, Phys. Rev. E. **57** 1145 (1998)

[94] A. Ishimaru, *Wave Propagation and scattering in Random Media volume 1 and 2* (Academis Press) (1978)

[95] R. Berkovitz, S. Feng, *Correlations in coherent Multiple scattering* Phys. Rep. **238**, 135 (1994)

[96] B. C. Brock, *Using vector Spherical harmonics to Compute Antenna Mutual Impedance from Measured or Computed Fields* (SANDIA Report SAND2000-2217-Revised) (2000)

[97] B. Altshuler, P. A. Lee, R. A. Webb *Mesoscopic Phenomena in Solids* (North-Holland, Amsterdam) (1991) Volume 30 of *Modern Problems in Condensed Matter Science*

[98] D. Vollhardt, P. Wölfle, *Selfconsistent theory of Anderson localization* in *Electronic phase transitions* Vol. 32 of *Modern Problems in Condensed Matter science* 1 (North-Holland, Amsterdam) (1992)

[99] T. Kirkpatrick, *Localization of acoustic waves*, Phys. Rev. B **31**, 5746 (1985)

[100] B. A. van Tiggelen, A. Lagendijk, A. Tip, *Speed of light in Random Media*, Phys. Rev. B **45**, 12233 (1992)

[101] C. T. Chan, K. M. Ho, C. M. Soukoulis, *A7 structure: a family of photonic crystals*, Phys. Rev. B **50**, 1988 (1994)

[102] C. T. Chan, K. M. Ho, C. M. Soukoulis, *Existence of a photonic gap in periodic dielectric structures*, Phys. Rev. Lett. **65**, 3152 (1990)

[103] H.A. Kramers, Nature, **117**, 775 (1926)

[104] R. de L. Kronig J. Opt. Soc. Am. **12**, 547 (1926)

[105] G.W. Milton, D J. Eyre, and J.V. Mantese, *Finite Frequency Range Kramers Kronig Relations: Bounds on the Dispersion* Phys. Rev. Lett. **79** (1997)

[106] Z. Zhang, P. Sheng, *Diffusion and and localization in random composites*, in *Scattering and localization of classical waves in random media* (World Scientific) (1990)

[107] Yu. N. Barabanenkov, L. M. Zurk, M. Yu. Barabanenkov, *Diffusion Asymptotics of the Bethe-Salpeter equation for elctromagnetic waves in dicrete random media* Physics Letters A **206**, 116 (1995)

[108] D. van Coevorden, *Light Propagation in ordered and disorederd media* Ph.D. Thesis, (Unioversity Amsterdam) (1997)

[109] D. Foster, *Hyarodynamic fluctuiations, broken symmetry, and correlation functions* (Addison-Wesley) (1990)

[110] F. MacKintosh, S. John, *Coherent backscattering of light in the presence of time-reversal-noninvariant and parity-nonconservibng media* Phys. Rev. B **37**, 1884 (1988)

[111] U. Fano, *Pressure broadening as a prototype of relaxation* Phys. Rev. **131**, 259 (1963)

[112] G. Papanicolaou, R. Burridge *Transport equations for the stokes parameter from Maxwell's equations in a random medium* J. Math. Phys. **16**, 2074 (1975)

[113] R. Loudon, *The quantum Theory of Light* (Clarendon, Oxford) (1983)

[114] A. Lagendijk, B. A. van Tiggelen, *Resonant multiple scattering of light*, Phys. Rep. C **270**, 143 (1996)

[115] B. A. van Tiggelen, A. Lagendijk, A. Tip, *Comment on: The proplem of light diffusion in strongly scattering media* Phys. Rev. Lett. **71**, 1284 (1993)

[116] U. Frisch, *Wave propagation in Random Media* in *Probabilistc Methods in Applied Mathematics Vol. 1* (Academic Press) (New York) (1968)

[117] G. Mahan, *Many-Particle Physics* (Plenum) (New York) (1981)

[118] C. F. Bohren, D. R. Huffman, *Absorption and Scattering of Light by Small Particles* (John Wiley and Sons, Inc) (1983)

[119] A. Gonis, *Green Functions for ordered and disordered systems (Studies in Mathematical Physics, Vol 4)* (Elsevier Science & Technology Books) (1992)

[120] K. Arya, Z. Su, J. L. Birman, *Anderson Localization of the Classical Electromagnetic Waves in a Disordered Dielectric Medium* in *Scattering and localization of classical waves in random media* (World Scientific) ed. by P. Sheng (1990)

[121] K. M. Case, P. F. Zweifel, *Linear transport theory (Addison-Wesley series in nuclear engineering)* (Addison-Wesley Pub.) (1967)

[122] J. Korringa, *Early History of multiple scattering theory for ordered systems*, Phys. Rep. **238**, 341 (1994)

[123] George Arfken, *Mathematical Methods for Physicists*, Academic Press, New York, (1968)

[124] Amic E, Luck JM, Nieuwenhuizen TM, *Anisotropic multiple scattering in diffusive media* J PHYS A-MATH GEN 29, 4915 AUG 21 (1996)

[125] Gevorkian ZS, Nieuwenhuizen TM, *Interference phenomena in radiation of a charged particle moving in a system with one-dimensional randomness* PHYS REV E 61, 4656, (2000)

[126] Luck JM, Nieuwenhuizen TM, *Light scattering from mesoscopic objects in diffusive media* EUR PHYS J B 7, 483, (1999)

[127] A. Lubatsch, *Propagation of Light and lasing Action in Disordered and Dielectric Media* Shaker Verlag, in print.

[128] A. Lubatsch, K. Busch, J. Kroha, *Transport of Light in Disordered and Laser Active Media* Phys. Rev. B, in preparation.

[129] van Rossum MCW, Nieuwenhuizen TM, *Multiple scattering of classical waves: microscopy, mesoscopy, and diffusion* REV MOD PHYS 71, 313 (1999)

[130] Nieuwenhuizen TM, van Duin CNA, *Theory of site-disordered magnets* EUR PHYS J B 7, 191 (1999)

[131] Lancaster D, Nieuwenhuizen TM, *Scattering from objects immersed in a diffusive medium* PHYSICA A 256, 417 (1998)

[132] Saakian DB, Nieuwenhuizen TM, *Variational approach to interfaces in random media: Negative variances and replica symmetry breaking* J PHYS I 7, 1513 (1997)

[133] Amic E, Luck JM, Nieuwenhuizen TM, *Multiple Rayleigh scattering of electromagnetic waves* J PHYS I 7, 445 (1997)

[134] Amic E, Luck JM, Nieuwenhuizen TM, *Anisotropic multiple scattering in diffusive media* J PHYS A 29, 4915, (1996)

[135] Mosk A, Nieuwenhuizen TM, Barnes C, *Theory of semiballistic wave propagation* PHYS REV B 53, 15914, (1996)

[136] vanTiggelen BA, Maynard R, Nieuwenhuizen TM, *Theory for multiple light scattering from Rayleigh scatterers in magnetic fields* PHYS REV E 53, 2881 (1996)

[137] VANROSSUM MCW, NIEUWENHUIZEN TM, HOFSTETTER E, *DENSITY-OF-STATES OF DISORDERED-SYSTEMS* PHYS REV B 49, 13377 (1994)

[138] NIEUWENHUIZEN TM, BURIN AL, KAGAN Y, *LIGHT-PROPAGATION IN A SOLID WITH RESONANT ATOMS AT RANDOM POSITIONS* PHYS LETT A 184, 360 (1994)

[139] NIEUWENHUIZEN TM, VANROSSUM MCW, *ROLE OF A SINGLE SCATTERER IN A MULTIPLE-SCATTERING MEDIUM* PHYS LETT A 177, 102 (1993)

[140] Scheffold F, *Particle sizing with diffusing wave spectroscopy* J DISPER SCI TECHNOL 23, 591 (2002)

[141] Rojas-Ochoa LF, Romer S, Scheffold F, *Diffusing wave spectroscopy and small-angle neutron scattering from concentrated colloidal suspensions* PHYS REV E 65, Art. No. 051403 Part 1 MAY 2002

[142] Garcia-Martin A, Scheffold F, Nieto-Vesperinas M, *Finite-size effects on intensity correlations in random media* Phys. Rev. Lett. **88**, 143901 (2002)

[143] Wyss HM, Romer S, Scheffold F, *Diffusing-wave spectroscopy of concentrated alumina suspensions during gelation* J Colloid. Interf. Sci. **241**, 89 (2001)

[144] Scheffold F, Lenke R, Tweer R, *Localization or classical diffusion of light?* Nature **398**, 206 (1999)

[145] Scheffold F, Maret G, *Universal conductance fluctuations of light* Phys. Rev. Lett. **81**, 5800 (1998)

[146] Scheffold F, Hartl W, Maret G, *Observation of long-range correlations in temporal intensity fluctuations of light* Phys. Rev. B **56**, 10942 (1997)

Acknowledgements

It is a great pleasure to acknowledge help from supervisors, colleagues and friends.
Foremost I would like to mention Prof. Hans Kroha, who guided me through the work of this thesis and was always a source of new ideas.
I also want to express my gratitude to Prof. Kurt Busch for many fruitful discussions and for refereeing this thesis.
Furthermore I want to thank Dr. Andreas Lubatsch for numerous, extenensive and sometimes thoroughgoing but productive discussions. It was a pleasure to work with him.
I would like to thank the groups of Prof. Hans Kroha, Prof. Hartmut Monien and Dr. Dmitry Chigrin for always being a source of help and discussion for any problem that arose. Special thanks go to the group of Prof. Kurt Busch.
Sincere thanks are given to my family, and I am very much indebted to all my friends.
Last but not least I want to thank two excellent - unfortunately retired - Profs. Franz Hasselbach and Walter Dittrich.

i want morebooks!

Buy your books fast and straightforward online - at one of world's fastest growing online book stores! Environmentally sound due to Print-on-Demand technologies.

Buy your books online at
www.get-morebooks.com

Kaufen Sie Ihre Bücher schnell und unkompliziert online – auf einer der am schnellsten wachsenden Buchhandelsplattformen weltweit! Dank Print-On-Demand umwelt- und ressourcenschonend produziert.

Bücher schneller online kaufen
www.morebooks.de

VDM Verlagsservicegesellschaft mbH
Heinrich-Böcking-Str. 6-8   Telefon: +49 681 3720 174   info@vdm-vsg.de
D - 66121 Saarbrücken   Telefax: +49 681 3720 1749   www.vdm-vsg.de

Printed by Books on Demand GmbH, Norderstedt / Germany